INTERACTIVE PHYSICS™
WORKBOOK

SECOND EDITION

CINDY SCHWARZ • JOHN ERTEL
Vassar College *U.S. Naval Academy*

WITH INTERACTIVE PHYSICS™ TEXTBOOK EDITION

Upper Saddle River, NJ 07458

Associate Editor: Christian Botting
Senior Editor: Erik Fahlgren
Editor-in-Chief, Science: John Challice
Vice President of Production & Manufacturing: David W. Riccardi
Executive Managing Editor: Kathleen Schiaparelli
Assistant Managing Editor: Becca Richter
Production Editor: Elizabeth Klug
Supplement Cover Manager: Paul Gourhan
Supplement Cover Designer: Joanne Alexandris
Manufacturing Buyer: Ilene Kahn
Editorial Assistant: Andrew Sobel
MSC.Software: Michael Woo
MSC.Software: Paul Mitiguy

Pearson Prentice Hall
Pearson Education, Inc.
Upper Saddle River, NJ 07458

Printed in the United States of America

10 9 8 7 6 5 4 3 2 1

ISBN 0-13-067108-8

Pearson Education Ltd., *London*
Pearson Education Australia Pty. Ltd., *Sydney*
Pearson Education Singapore, Pte. Ltd.
Pearson Education North Asia Ltd., *Hong Kong*
Pearson Education Canada, Inc., *Toronto*
Pearson Educación de Mexico, S.A. de C.V.
Pearson Education—Japan, *Tokyo*
Pearson Education Malaysia, Pte. Ltd.

About the Authors

Dr. Cindy Schwarz is currently Associate Professor of Physics at Vassar College in Poughkeepsie, NY. She majored in physics and mathematics at S.U.N.Y. Binghamton, graduating in 1980. She received her Ph.D. from Yale University in 1985 for her work in experimental particle physics. She has written two other books *A Tour of the Subatomic Zoo: A Guide to Particle Physics* (Springer-Verlag) and *Tales from the Subatomic Zoo* (self-published at www.smallworldbooks.net) and a multimedia CD-ROM *Interactive Journey through Physics* (Prentice Hall). She is a member of the American Physical Society (APS), the American Association of Physics Teachers (AAPT), and is a member of the Undergraduate Texts in Contemporary Physics (UTCP) Editorial Board for Springer-Verlag. Cindy and John first collaborated at AAPT national meetings, giving workshops on beginning through advanced level *Interactive Physics*™. Cindy has received two grants from the National Science Foundation (NSF) for work in curriculum development using technology to aid in the teaching of physics. Most recently, she was awarded a grant in the NSF Course Curriculum and Laboratory Improvement (CCLI) program for development of a course for non-majors using digital video and analysis software to teach and assist in the learning of physics. Cindy is married and lives in Staatsburg New York with her husband Norman Rachmilowitz, her sons Michael and Bryan, and their dog Kellie.

Dr. John Ertel is Associate Professor of Physics at the Unites States Naval Academy in Annapolis, MD. He received his BS in 1966 and his MS in 1968 from Emory University for his theoretical work describing the angular correlation of alpha particles emitted in the spontaneous fission of Californium-252. In the summer of 1968, John went on active duty with the Marine Corps where he served as a pilot. While still in the Marines, he started part-time work on his Ph.D. After exiting active duty service, John was invited to return to the Academy as an Assistant Professor while continuing work towards the doctorate. In 1983, Professor Ertel received his Ph.D. from the Catholic University of America in Washington, D.C. for his work in theoretical electron scattering. He then returned as a faculty member of the Physics Department of the U.S. Naval Academy where he has remained since that time. Since the debut of *Interactive Physics* in the late 1980's, Professor Ertel has given workshops on the simulation software at over twenty meetings of the American Association of Physics Teachers as well as numerous colleges and universities. In the summer of 1997, Dr. Schwarz added her expertise to these long running workshop presentations. Dr. Ertel is an active member of APS, Acoustical Society of America (ASA), and AAPT and is a member of the UTCP Editorial Board for Springer-Verlag. In his local community, he is currently the Governor's Select Appointee to the Board of Trustees, Maryland School for the Deaf. John is also very active in the Boy Scouts of America. He lives in Annapolis with his wife, Patricia Burt, and their dogs Maggie and Cinders.

Contents

Author Comments on the First Edition

Motion is what much of physics is about. We have equations that we can solve for velocity, acceleration, energy, and more. We can think about the motion, analyze it, and try to understand it, but until very recently we could not "see" it demonstrated so clearly on the computer. Traditional physics courses consist of lectures where concepts are introduced by the professor, equations are derived, and example problems are solved for the student. I am a product of that traditional course myself. In 1976, my freshman year in college, I took introductory physics. I sat way in the back of the very large lecture hall, listening passively to the professor describe the motion of a projectile, wondering what it looked like. So, naturally, in 1985, my first year teaching at Vassar, I stood in front of 40 students trying to explain the motion of a block attached to a spring on a frictionless surface. I drew vectors for acceleration and velocity at representative times on the blackboard. Snapshots you might call them, but it was static, it was not moving, and motion was what I was trying to explain.

Fortunately, many courses are now changing to incorporate new techniques including cooperative learning, more hands-on experiences, and the use of computers. The computer is being used in many schools for data collection and analysis, and running simulation software. *Interactive Physics* (IP), first introduced in 1989, is now one of the best simulation programs around, adding a new dimension to the way physics is taught and offering a place for students (and teachers) to explore motion in ways they never could before. In 1989 I made my first attempts at using *Interactive Physics* for demonstrations in the classroom. We looked at a ball launched into the air and its subsequent motion. My students and I could see the motion and see the velocity vector of the ball. We played "what if" games. We changed the angle that the ball was launched with and saw how the trajectory of the ball changed. We asked ourselves, "Will it go higher if the ball weighed more?" "What is the velocity of the ball at the top of the flight?" "Which will go farther, a steel ball or a Ping-Pong ball?" We got answers to all those questions. This was certainly a step in the right direction.

I continued to use *Interactive Physics* with my students, although I started getting them involved in designing their own simulations. This was useful, but time-consuming. Meanwhile, some of the publishers (and authors) of introductory physics texts were also realizing that *Interactive Physics* could be very useful. Several texts now have end-of-chapter problems correlated with *Interactive Physics* simulation files. This I also felt was a step in the right direction, but it was sort of backward. The problems were being "translated" into *Interactive Physics*. It was good to be able to visualize the motion for these problems, but some of the best features of *Interactive Physics* were not being fully exploited. So, in 1994, as I wrote this book, I tried to find the best ways to use the technology. I tried to cover the major areas in mechanics and address what I know to be (from 10 years of teaching) some of the most common student misconceptions. Now you can make the objects collide and stick together or collide and bounce apart, you can see the motion, simple or complex. You can play "what if" games and get answers to your questions.

This workbook is intended to be a supplement to any algebra- or calculus-based introductory physics textbook (HS or College Level). It is an interactive workbook, designed exclusively to be used with the simulations that I created using *Interactive Physics*. I know that this workbook and the 40 simulation files that go with it will help you to understand physics better. I hope that it will also be fun. It has now been almost 20 years since I sat in that lecture hall in my introductory physics course at S.U.N.Y. Binghamton. Many things have changed. We used computers rarely and only for writing programs on mainframe machines. We did not have computers in our introductory physics labs, our classrooms, our dormitories, or our homes. We did not have the technology to visualize the motion. We saw films, we tried to see things, but we were limited. You are not. Take advantage of the window that *Interactive Physics* and this workbook gives you and look through to see, explore, and understand motion and physics better than ever.

Cindy Schwarz 1996

Author Comments on the Second Edition

The preface to the first edition opens with the wonderfully simple statement "Motion is what much of physics is about." Following the lead of this statement, in the following paragraphs, I would like to describe the typical uses for which we at the Naval Academy have found *Interactive Physics* to be an invaluable pedagogical aid to learning and understanding for my students (and sometimes, myself).

Classroom Use of IP—Stage-I (Baiting the Hook)

In the fall term of introductory physics, most of our classes concentrate on the area of Mechanics. At the very beginning of the term, IP is introduced by modeling the simplest of kinematic problems in class before the students. By using IP right from the start, we allow/force the student to visualize the motion associated with what may seem to them to be very dry algebraic or calculus equations. In each problem attempted in class, the student is shown:

• how easy it is to "program a simulation" of that motion using the syntax of IP. Translated to common language, this means you draw a picture of the experiment using the graphical user interface of IP and the simulation is ready to run.
• how closely the numbers extracted from "the motion simulated in IP" matches their own calculations or those done by their teacher or in their text
• when their calculations differ from the motion we simulated in IP, that the motion they see in IP makes more physical sense and is therefore more likely to be correct
• when their own calculations differ from the motion we simulated in IP, how they may use the simulation to find errors in their calculations

In this first stage we guide the students toward their "I believe button" and start building their confidence both in IP and in the proper mathematical description of mechanical processes.

Classroom Use of IP—Stage-II (Setting the Hook)

After a very short period of time (sometimes almost immediately), the student begins to ask questions like "What happens to the motion if you increase the value of 'this' or decrease the value of 'that'?" Once you get to this point with any portion of the class, the effect tends to snowball dragging the rest along. Here I find the first use of "small problem-solving groups" of three or four interacting among themselves and determining how they think the motion in an experiment will change when certain parameters are varied. This leads very naturally to cross-group discussion about why their particular group answer makes more sense than other groups', which immediately leads to the almost demanding question "Could you change the acceleration [or some other variable] so we can see what actually happens?" Now you almost have their complete and voluntary attention!

Classroom Use of IP—Stage-III (Reeling Them IN)
In the last classroom stage, as their homework may seem to them to grow in complexity, we can offer them the opportunity to "play with the software" and write their own IP simulations on our departmental computers.*

Laboratory Use of IP—(Preparing and Enjoying the Feast)
As a capstone to this evolution towards "guided discovery", once the student becomes familiar with using IP as a partner in the study of motion, before each scheduled laboratory, we allow them to simulate each and every experiment using the IP simulation environment on our laboratory workstations. The importance of using "real world" values for parameters in these simulations cannot be over emphasized. As an example, if you are studying *Hooke's Law and Spring-Mass Oscillations* using one of the large conical brass springs that are commercially available, then use a spring constant of about 10 N/m in the IP simulation of the motion, and the visualization they will see will seem all the more real. By the way, the student that does a good job of taking data will notice that their experimental values and those from the IP simulation will slightly disagree—as if the mass value put into IP should be increased. Well, this can be immediate evidence to the student that the mass of the spring is actually an integral part of the kinematics of the motion, since at least part of the spring is moving as well.

In closing the preface to the first edition, Cindy wrote "This workbook is intended to be a supplement to any algebra- or calculus-based introductory physics textbook (HS or College Level). It is an interactive workbook . . . Take advantage of the window that *Interactive Physics* and this workbook gives you and look through to see, explore, and understand motion and physics better than ever." The use of *Interactive Physics* in both the classroom and in the laboratory leads the student along exactly that suggested path.

Now you really have them—they have been enticed into the holy grail of all education, being actively engaged in the process of the growth of their own knowledge and understanding. No longer passively trying to memorize the "gouge equations", now they can be "self-starters" in the learning of physics.

If it sounds as if I argue too strongly for the merits of using *Interactive Physics* as an adjunct in teaching mechanics, it is only because I have found it to be such a great help to my students. I certainly don't want to become a salesman for any particular piece of software! But then, is this really any different from choosing the textbook our students must use?

In this workbook we offer students of physics and engineering a helpful learning aid, including a "textbook version" of *Interactive Physics*. Once they have found out how useful this workbook and associated software is in easing the pain some feel while being subjected to Introductory Physics, they may well want to pursue an upgrade to the full version of *Interactive Physics* and find further enjoyment in understanding the physical universe in which we all live.

Have fun with this workbook, I know we certainly did in creating it!

John Ertel 2003

---

* At the Naval Academy, we teach approximately 1200 students per term with the necessity of keeping the class size down to a maximum of twenty-four. This necessitates that we have approximately seven laboratories each of which has eight computer workstations. By opening these laboratories up to our students in the off duty hours, we allow them work in the IP environment almost anytime they please.

Well, how much more can I add. We really want you to get to the simulations, so a few more words must suffice. Much has changed since the first edition of this book was written almost 10 years ago. The computers we have are faster and the access to them is much more universal. Visualization has become an important tool in physics and many other fields of study. In particular, 3-D visualization has become used extensively in physics in research and to a lesser extent in the classroom. Although the simulations included in this workbook are based on technology that has been around for a while and they are not state of the art in computer animation and visualization, they are nevertheless extremely useful as a teaching and learning environment in the physics classroom. In preparing this second edition John and I have maintained the features of the book that were the most useful to our students and the students and teachers who used the first edition with much enjoyment and success, while modifying some to include new graphics and features. The software has evolved through several improved versions initially at Knowledge Revolution and then at MSC.Software and without the *Interactive Physics* program this book would not even exist. We are grateful to many people at MSC.Software, and Prentice Hall for their assistance on this project and for those of you who have used the book and provided valuable feedback. We thank our families for their patience throughout the process. I thank my husband, Norman, and my two sons, Michael and Bryan, and I know John's wife, Patricia Burt, has been a behind-the-scenes helper.

Cindy Schwarz Rachmilowitz 2003

Layout of the Book

There are 40 simulation exercises in the workbook. Although you will find some exceptions, generally each has the following four sections:

- Physics Review
- Simulation Details
- The Exploration
- Self-Test Questions

In the physics review section you will find a brief review of the physics concepts covered in the simulation. Depending on the text that you are using, the symbols and terminology used may or may not be exactly what you are familiar with, but they should be similar. This section is not intended to replace your textbook, but rather to review and summarize concepts that you should already have covered, that are necessary to complete the simulation. If you encounter something that you do not understand, you should look in your text or see your instructor. Reference topics (listed at the end of the physics review) suggest key words to look up. This section also will contain formulas that you should be familiar with.

In the simulation details section you will find most of the information necessary to run each simulation. You will be told what parameters can be adjusted and the fixed values of those that cannot be adjusted. You should read this section carefully before beginning the simulation and then refer back to it as needed. There are some techniques that are common to all simulations and these are described in "The *Interactive Physics* Environment" in the next section.

The exploration section is the area where you do most of the work. Each simulation is different, but some of the most common tasks are:

Doing Things on Your Computer Screen—To make it easier for you to identify when you must do things in the *Interactive Physics* environment, there are gray background areas that start out with an icon of a mouse, and what follows are all instructions pertaining to the computer simulation: clicking on buttons, adjusting inputs with a slider, turning on air resistance, and so forth.

Recording Simulation Data into the Workbook—Sometimes it will simply be a matter of recording a number shown in the simulation data explicitly, and sometimes you will have to manipulate the data shown to calculate a quantity that is not shown by a meter. The icon will remind you to use the tape player. For more on how to use the tape player, see pages xiv–xv.

Sketching Graphs—In some cases you will be sketching the results shown in the simulation so that you can have them for comparison to another case or so that you can identify key features. Other times you will be predicting what a graph will look like if you change the value of some parameter. What is important is the general nature and shape of the graph.

Predicting What Will Change—If you can change the value of a parameter, you will be asked to predict how it will (or will not) affect other outputs in the simulation. Predicting correctly is not as important as predicting and understanding the agreement (or disagreement) of your prediction with the actual results.

Plotting Simulation Data—When more exact graphs are required, you will be asked to plot the simulation data. Often you will be given a grid with numbers already marked. You can plot by hand or use a spreadsheet or graphing program if you have access to one.

Doing Calculations—You will have to calculate something that may be compared to a simulation value or used as an input parameter for the simulation.

Entering Data into a Table—All tables have titles and are sectioned off by double lines. Some of the entries may be data shown explicitly in the simulation and others may have to be calculated.

Answering Questions—All questions are in italics. You are encouraged to discuss your answers with other students in your class or your instructor.

Each simulation (except the last five) will end with a self-test. These questions should be answered **only after you have finished the simulation exercises**. The questions will be either true/false or short answer, and the answers to all of them are in the back of the book. Our recommendation is that if you get more than one wrong, you should repeat part or all of that simulation before going on to the next one. The answers to all of the questions in simulations 36, 37, 38 and 39 are in the back of the book as well.

Other boxed sections that you will find in many of the simulations include:

Hint—Some calculations and predictions may be more difficult to do. These boxes can help get you off on the right track. You can use them or skip them as you like.

Explanation—Where more explanation of the results and the physics was needed, these boxes were included. They go into more details specific to the simulation than the physics review sections.

Mathematics Help—These boxes will review some mathematics or contain information to make calculations easier.

Optional—There are some optional discussions, derivations, and graphing exercises that can be omitted or included as you (or your instructor) see fit.

A Note about Notation and Formulas

We tried to be consistent with the symbols used in *Interactive Physics* and we used ones that were the most common. However, the symbols may not always agree with what your textbook or instructor uses. For example, *Interactive Physics* and this workbook use FF to represent the force of friction (you may know it as F_f) and FN to represent the normal force (you may know it as F_N). Most textbooks use a for acceleration, but *Interactive Physics* uses A. We have tried to define all relevant symbols in the physics review section, to eliminate confusion. When in doubt, use your "physics sense."

Warnings and Tips!

You should be very careful to reset the simulations each time before changing values on the sliders or in the input boxes. If you forget to reset and you change the value of something, the results can be unpredictable. The worst that can happen is that all of the objects on the screen will disappear and in the best case, it will be okay. So to be on the safe side, **ALWAYS RESET BEFORE CHANGING ANY VALUES UNLESS SPECIFICALLY TOLD OTHERWISE.**

One thing that can definitely cause problems is typing a number into a box on the screen without resetting. So to be on the safe side, always use the slider to change values unless specifically told otherwise. Never fear though, if you do something to mess up the simulation, simply close the file and then reopen it again and all should be back to normal. If you are asked to save changes, **CLICK ON "NO"!**

The retain/erase graph button is a toggle button. Unfortunately, there is no way to indicate whether it is set to retain graphs or erase them. If you are uncertain of its state, click it once and run. If the desired result is not obtained, you should click again.

Software and System Requirements
The simulation files for this workbook come with the software needed to run them as they are. These files can also be run with the full version of *Interactive Physics*. If you own or have access to the full version, you will be able to create your own simulations.

Macintosh
- PowerPC-based system
- MacOS System 7.1 or above; the classic mode will be used for MacOS X machines
- 32 MB of physical RAM
- 60 MB hard disk space
- CD-ROM Drive
- Monitor (256 colors with 800 x 600 resolution or higher)
- Mouse

Windows-based Systems
- Windows 98 or higher operating system
- Pentium-based PC or equivalent
- 16 MB RAM Minimum
- 60 MB hard disk space
- CD-ROM Drive
- Monitor (256 colors with 800 x 600 resolution or higher)
- Mouse

Installing and Running FILES
Before you can run the simulations that go along with the text, you must install them and the *Interactive Physics* Textbook Edition onto your computer. Detailed instructions can be found in the *Read Me First!* File for the Macintosh and in the *readme.txt* file for Windows-based systems.

Sliders, Buttons and Input Boxes
Real-time input devices in the *Interactive Physics* environment include sliders, buttons, and text fields. For example, some of the input sliders may control initial velocity or acceleration, mass, spring constant and so forth. The following explains how to adjust an angle (measured in degrees). To change the value, one must hold the mouse button down on the vertical bar and, while holding, slide left or right to the value you desire. Then release the mouse button. You may also, after resetting, type a number into the box. The number must be within the range of the slider. You can type in 24.23 into the box, and although you will see 24.2 displayed, 24.23 is the number used in the simulation to do calculations.

angle

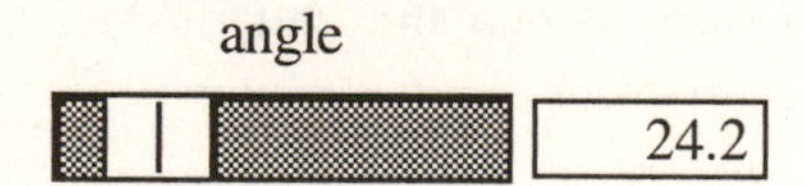

Buttons can perform a variety of functions. When you click on the Jogger button shown below with your mouse in one of the simulations, you will change the reference frame to that of a jogger running by a river.

Jogger

Text boxes allow you to type in a number within a predetermined range. This one will place a value of 30.00 m/s as the initial speed of an object.

initial speed

30.00

Output Graphs, Meters, and Vectors
Real-time output devices include graphs, digital displays, and bar displays. You may see meters for the position of a ball, the kinetic energy of a car, the force on a block, and so forth. Meters can display information in the physics world as:

- numbers (digits)
- level indicator (bars)
- graphs (with time as the *x* axis)

Vectors, which are used to represent the properties of velocity, acceleration, and force, are shown as arrows. The direction of the arrow shows the direction of the velocity, acceleration, or force. The length of the arrow corresponds to the magnitude of the velocity, acceleration, or force. Vectors may be shown as having magnitude and direction and/or in component (*x*, *y*) form.

Using the Tape Player Controls
While running a simulation, *Interactive Physics* also records it, using a feature called the tape player. This allows you to play simulations backward, to skip frames of the simulation, and to play simulations more quickly after all calculations have been completed. You will have to use the tape player a lot to record data and study what is happening in each simulation. In most cases you should run the simulation until it stops and then use the tape player to go back to the places that you need to. The tape player controls also provide a visual indication of the number of frames recorded.

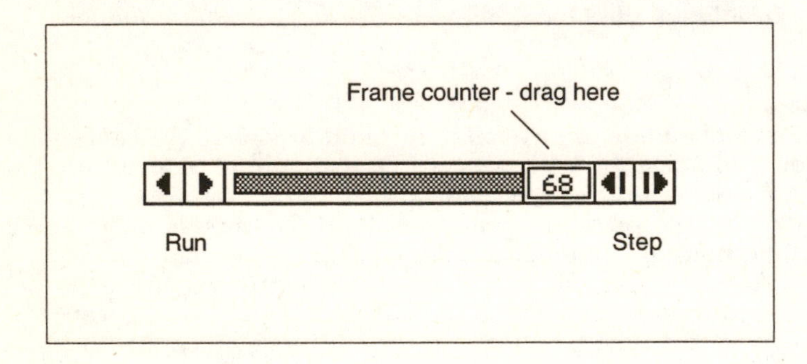

Stepping through frames
Interactive Physics allows you to view the recording of a simulation frame by frame.

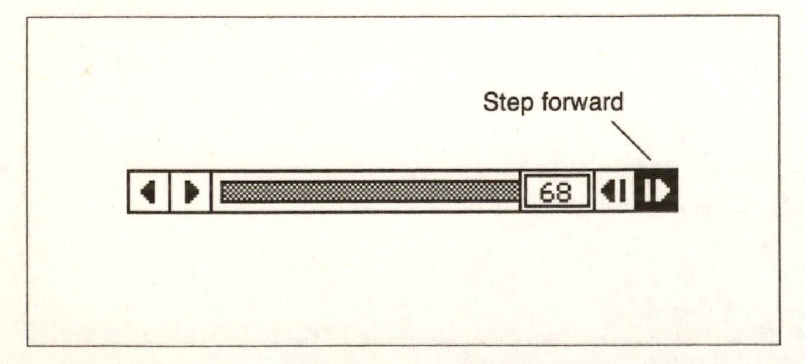

To step through a simulation click on the Forward or Backward step in the tape player control to move forward or backward one frame at a time. As you step through, all output devices show the values that were recorded in that frame and all vectors will show as they were in that frame.

Playing a simulation backward
After you have run a simulation, you can play it backward. Click the Run Backward control on the tape player. The simulation begins running in reverse. You can stop the simulation at any time. The simulation will stop by itself when it reaches the first frame.

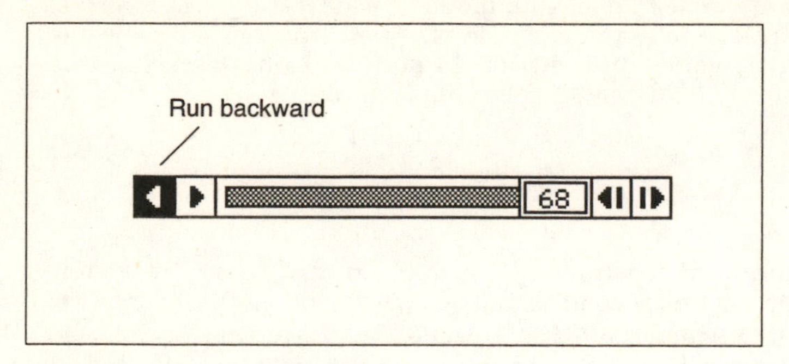

Moving to a specific frame
To move quickly to any frame in a simulation, drag the frame indicator left or right.

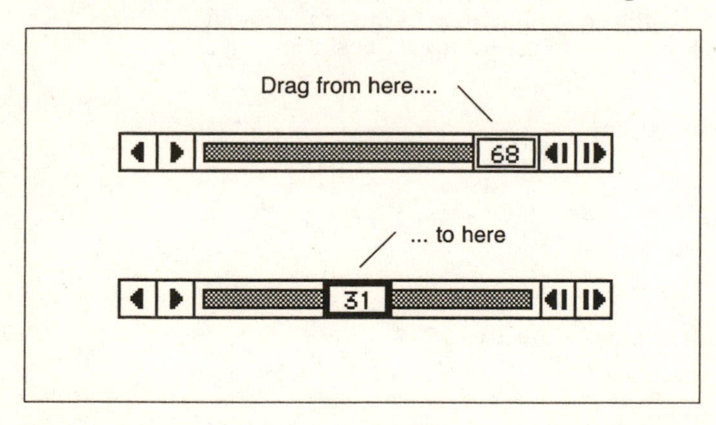

You can also click in any portion of the gray region on the tape player controls to immediately move the frame indicator to that location. To continue the simulation beyond the current recording, drag the frame indicator as far to the right as it will go and then click the Run Forward control. *Interactive Physics* now continues the simulation.

Speeding up playback
When you run a simulation for the first time, *Interactive Physics* not only draws the animation on the screen, but also calculates the physics that generates the animation. For most simulations this does not noticeably slow down the animation since *Interactive Physics* quickly calculates the physics. For complicated simulations, particularly those with many objects touching each other at the same time, the animation may be slow the first time you run the simulation. You can speed up the animation by playing the simulation again.

Replaying the simulation
To replay the recorded frames for faster animation:

- Record the animation and calculate the physics by running the simulation once.
- Click the Reset button.
- Run the simulation again.

The animation replays faster this time because *Interactive Physics* did not have to calculate the physics while replaying the simulation.

How Does *Interactive Physics* Work?
It is not necessary to understand how *Interactive Physics* works to be able to use it effectively, but for those of you who are interested, these are some of the basics. *Interactive Physics* solves problems by using numerical techniques. A systematic approach is used, which allows *Interactive Physics* to model a wide variety of problems. In describing the solution process, let's take the example of a ball traveling in projectile motion, with only the force of gravity acting on the ball. In standard physics texts, this problem is typically solved analytically by using the following formulas:

$$x = x_o + v_{ox}t$$
$$y = y_o + v_{oy}t - \tfrac{1}{2}gt^2$$

With these formulas one can find the position of the ball at any time by simply substituting the correct values for the initial conditions (x_o, y_o, v_{ox} and v_{oy}). If projectile motion was the only type of problem *Interactive Physics* needed to solve, this would also be an acceptable way to proceed on the computer. As problems get more complicated, it becomes increasingly difficult to solve the equations that describe the motion.

What equations are common to many physics problems? In our universe, neglecting relativity, some standard equations and definitions in mechanics are:

$$F = ma$$
(Force = mass x acceleration)

$$\tau = I\alpha$$
(Torque = moment of inertia x angular acceleration)

$$a = \frac{dv}{dt}$$
(instantaneous acceleration = derivative of v with respect to t)

$$v = \frac{dx}{dt}$$
(instantaneous velocity = derivative of x with respect to t)

These are the equations that *Interactive Physics* uses in its solution of dynamics problems. The general solution is typically carried out by a process known as integration. If $a = dv/dt$, then v = integral of $a*dt$. The process of integrating on a computer is similar to predicting, and then checking to see how good your prediction is. What *Interactive Physics* does is to find the current acceleration of an object and use this acceleration to derive a new velocity at some small time later. This process is then used again to formulate a new position. In formulas, the process looks something like this:

Now at $t = 0$. Calculate a.

Use a to calculate v at $t = 1$.

$v(t = 1) = v(t = 0) + a*dt$

Use v to calculate x at $t = 1$.

$x(t = 1) = x(t = 0) + v*dt$

This whole process is called numerical integration. There are many methods for using an acceleration to compute new velocity and position terms at some later time. The one shown above is known as Euler integration and is the simplest type available. There are two other methods that *Interactive Physics* can use depending on whether accuracy or speed is more important.

An important thing to realize about solutions obtained with numerical integrators is that they don't exactly match the solutions one would obtain with the analytic (textbook formula) solution. The reason the solution doesn't absolutely match the formula solution resides in the method used by *Interactive Physics* to obtain results. Integration breaks the problem down into several small steps. Each step involves finding the acceleration of all objects, and then formulating (predicting) what the velocity and position of objects should be at a short time in the future. The quality of this method is intimately tied to how big the time intervals are that are used to step to new velocities and accelerations. The size of these intervals is known as the time step. The time step is a critical variable in all simulations based on numerical integration. As you might imagine, using smaller time steps results in better predictions for the new velocities and positions that are generated as the integrator steps along. In general, a smaller time step produces more accurate simulation results. Using small time steps means that the computer will have to make more calculations to simulate a certain time (a second, for example) of animation.

Thus, the classical trade off:

Small time step = good accuracy

Large time step = good computing speed

The problem of picking a time step is difficult! *Interactive Physics* does its best to shield you from the decision of picking a time step, unless you decide that you want to set it yourself (which you cannot do with the version of IP included). Some of the simulations in this workbook had to be done with small time steps to provide accurate numbers for doing precise calculations, and therefore they may run more slowly on your computer. For other simulations, where it was more important to get quick animation rather than exact numerical values, longer time steps were used. You may notice that in some cases your calculated values do not agree exactly with the simulation data, but they should be very close. For example, if the initial kinetic energy of a block is 50.00 J prior to a collision and the simulation shows 50.07 for the kinetic energy of the block after the collision, one should conclude that energy has been conserved, because for all intents and purposes, these two numbers are equal. However if the kinetic energy of the block after the collision was 47.25 one could conclude that some energy was lost. You may have to use your judgment and some "physics intuition" to make these decisions. You may also notice that many simulations do not stop exactly when an object has arrived at some location, or exactly when its velocity is 0.00. That is also because of the chosen time step. The simulations can be made to stop close to those points, and the more accurate that is, the shorter the time step and the slower the simulation will be. In some cases it would be painfully slow to get a more accurate simulation. Hence, again the trade-off. In designing each simulation, we tried to make the best choices to get the right balance between accuracy and speed.

A Note For Instructors
Although the physics review sections at the beginning of each simulation suggests reference topics, which should clue you in to what topics are covered, the following table contains some additional information that can be useful as well. The first part of the table separates the simulations by level. Level I simulations are the most basic. They will help those students who are having difficulty most, but should be useful for all students. The level II simulations tend to include more than one topic and should be useful for all students. The level III simulations are the most challenging. All of the simulations can be done without knowledge of calculus, but two of them have optional calculus derivations included (14 and 33). The second section of the table gives the type of pedagogy that is emphasized in the simulations listed. The third section breaks things down by topic. Note, of course, that several problems are listed under more than one topic. Note that several problems are listed under more than one topic.

Table for Instructors

Level I	1, 2, 3, 8, 16, 22, 25, 26, 28, 29
Level II	4, 5, 6, 7, 9, 10, 11, 12, 13, 14, 15, 17, 19, 20, 24, 27, 30, 31, 35, 39, 40
Level III	18, 21, 23, 32, 33, 34, 36, 37, 38,39

Predicting and Understanding Graphs	2, 3, 8, 10, 18, 19, 23, 30
Optional Spreadsheet or Graphing Program	8, 13, 15, 18, 24
Derivations and Calculations	14, 15, 18, 20, 24, 33, 34
Understanding Vectors	5, 6, 12, 24, 31

One Dimensional Motion	1, 2, 3, 4
Vector Addition	5, 6, 12
Free Fall	7, 8
Inclined Plane	16, 17, 18, 19, 39
Friction	13, 14, 15, 17, 18, 19, 20, 21, 32, 36, 38
Energy	13, 14, 18, 21, 22, 23, 24, 32, 35, 36, 37, 38, 39
Collisions	22, 24, 37, 38, 39
Projectile Motion	9, 10, 11, 23, 35
Simple Harmonic Motion	31, 32, 33, 38
Circular Motion	25, 26, 27, 29, 30
Rolling	34, 39
Equilibrium	28

The last five simulations do not have self-test questions. Simulations 36, 37 and 38 and 39 are more free form than the rest of the simulations and they integrate several topics. Simulation 40 is included to show the student more complex motion and therefore it has no quantitative questions. Remember that all of these simulations can be run with the software that comes with this book.

SIMULATION 1 A CAR MOVING WITH CONSTANT ACCELERATION (VECTORS)

Physics Review

Average velocity is defined as the change in an object's displacement in a time interval (Δt), and average acceleration is in turn defined as the change in an object's velocity in a time Δt (see formulas below). As the time interval is made smaller and smaller, these average values approach what we call the instantaneous values (i.e., the velocity of the car at 3 seconds is 10 m/s). Velocity and acceleration are vector quantities specified by a magnitude (size) and direction. Speed is the same thing as the magnitude of the velocity. We will start by studying the motion of a car moving in one dimension (i.e., it can only go back and forth along a straight road) with constant acceleration. In this special case the acceleration and velocity of the car will have a magnitude (e.g., 2 m/s^2 and 4 m/s) and the direction will be specified by the sign (+ indicates one direction and – indicates the opposite direction). When we reduce things to one-dimensional motion with constant acceleration, we can use the scalar equation (shown below) to solve for the instantaneous velocity (at a time t) given the acceleration and the initial velocity.

Formulas that you should be familiar with:

$$\vec{v} = \frac{\Delta\vec{x}}{\Delta t} \qquad \vec{a} = \frac{\Delta\vec{v}}{\Delta t} \qquad v = v_o + at$$

Reference topics: one-dimensional motion, velocity, acceleration, kinematics

Simulation Details

The car has an initial velocity that you can adjust with a slider (+ values will give vectors pointing to the right). You can also use a slider to set the acceleration of the car to different values and predict and observe the resulting motion. When you run this simulation, you will see the vectors representing the magnitude and direction of the velocity and acceleration of the car. You can stop the simulation at any time and then either click Run to continue from that point or Reset to initial conditions. To assure that you can see the car (and hence the vectors) at all times, click on the button labeled Car, which will put you in the frame of reference of the car. Then when you view the simulation, it will be as if you are riding in the car. To view the simulation as if you were a passerby on the side of the road, click on the button Ground to go to that reference frame. The simulation will automatically stop after 12.25 seconds have elapsed.

Click on **Car** to set the reference frame to that of the car. Set the initial velocity of the car to any positive value. Set the acceleration of the car to a positive value. Click **Run** and observe the motion of the car. You can reset and run again as many times as you like or stop and continue running from that point.

Is the acceleration vector pointing to the left or the right? Is its size changing?

Is the velocity vector pointing to the left or the right? Is its size changing? Why?

Click **Reset**. Set the initial velocity to 2 m/s and the acceleration to -0.30 m/s^2. Click **Ground** to view the simulation from the side of the road. Click **Run** and observe the motion of the car.

Is the acceleration vector pointing to the left or the right? Is its size changing?

Is the velocity vector pointing to the left or the right? Is its size changing?

Does the car ever come (momentarily) to a stop? (i.e., is its velocity ever zero?)

Once the car is moving to the left, is the magnitude of the velocity increasing or decreasing?

Will the car also reverse its direction if the initial velocity is negative and the acceleration is positive?

Let's test and see.

Click **Reset**. Set the initial velocity to –2 m/s and the acceleration to 0.40 m/s^2. Click **Run** and observe until the simulation stops.

Did the car reverse direction?

Explanation

You should now have seen that the car will reverse direction only if the initial velocity and acceleration vectors are opposite. It does not matter which is positive and which is negative. But will it always turn around? No, not if it runs out of time and the simulation automatically stops.

Use the [tape icon] to find the following:

the net displacement of the car = ________________

the total distance traveled by the car = ___________

Can you describe the difference between these two things?

Which car do you think will travel farther?

Car A—initial velocity of 4 m/s and acceleration of 0.5 m/s^2.

Car B—initial velocity of 2 m/s and acceleration of 1.0 m/s^2.

Click **Reset**. Try each of the above cases, running each until the simulation stops.

From the simulation data, record the position of each car.

position of car A = _________

position of car B = _________

Self-Test Questions for Simulation 1

These statements all apply to a car moving in one dimension. True or false?

1. A car with positive velocity that is increasing in magnitude must have a positive acceleration.
2. A car that has a negative acceleration must be slowing down.
3. A car that has a velocity and acceleration in opposite directions is slowing down.
4. A car that has a constant velocity must have zero acceleration.
5. If a car reverses direction, there must be an acceleration opposite to its original velocity.

SIMULATION 2 A CAR MOVING WITH CONSTANT ACCELERATION (GRAPHS)

Physics Review

For a car moving with a constant acceleration in one dimension, the velocity of the car can be calculated at any time *t*, using the equation below. In the previous simulation you studied the relationship between the acceleration and velocity vectors. In this simulation you will look at the graphs of velocity and acceleration as functions of time. Since the motion is only either to the left or the right, we can use scalar (rather than vector) notation.

Formulas that you should be familiar with:

$$a = \frac{\Delta v}{\Delta t} \qquad v = v_o + at$$

Reference topics: one-dimensional motion, velocity, acceleration, kinematics

Simulation Details

As in the previous simulation, the car has an adjustable initial velocity and acceleration. Since the motion is in one dimension, the vectors that point to the right will be positive numbers when graphed and the vectors that point to the left will be negative numbers. Pay attention to the graphs as the motion progresses and remember that once you have run a simulation, you can use the tape player to go back to any point that you want and study it in more detail. If you cannot recall how to do this and are having trouble, go back and read the section in the preface on "Using the Tape Player." The simulation will automatically stop after 10 seconds have elapsed. The initial position of the car is $x = 0.0$ m.

Set the initial velocity of the car to any positive value. Set the acceleration of the car to zero. Click Run and observe the motion of the car.

Sketch the velocity and acceleration graphs as shown in the simulation.

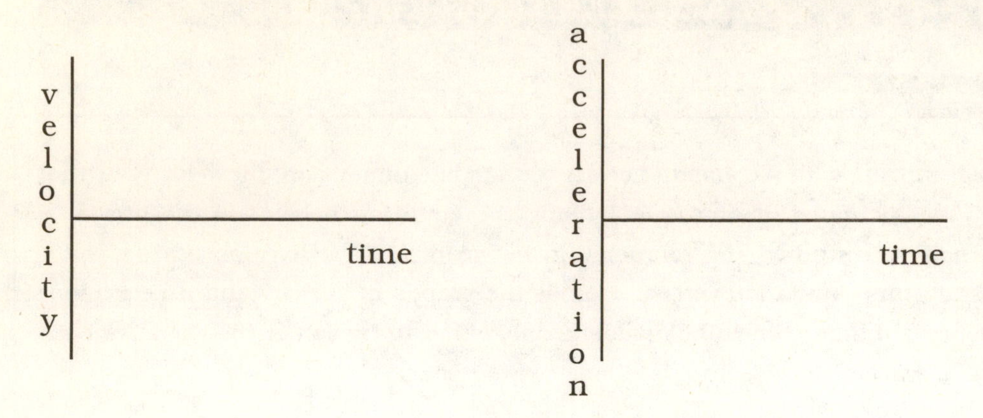

How do the graphs correlate with what is happening to the vectors for velocity and acceleration?

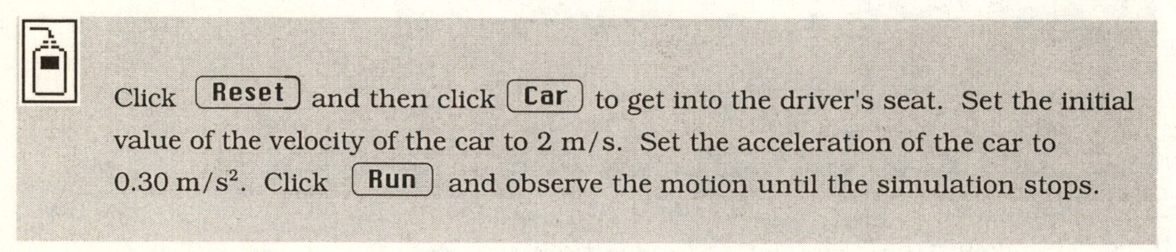

Click Reset and then click Car to get into the driver's seat. Set the initial value of the velocity of the car to 2 m/s. Set the acceleration of the car to 0.30 m/s^2. Click Run and observe the motion until the simulation stops.

Sketch the velocity and acceleration graphs, shown in the simulation, in the space below.

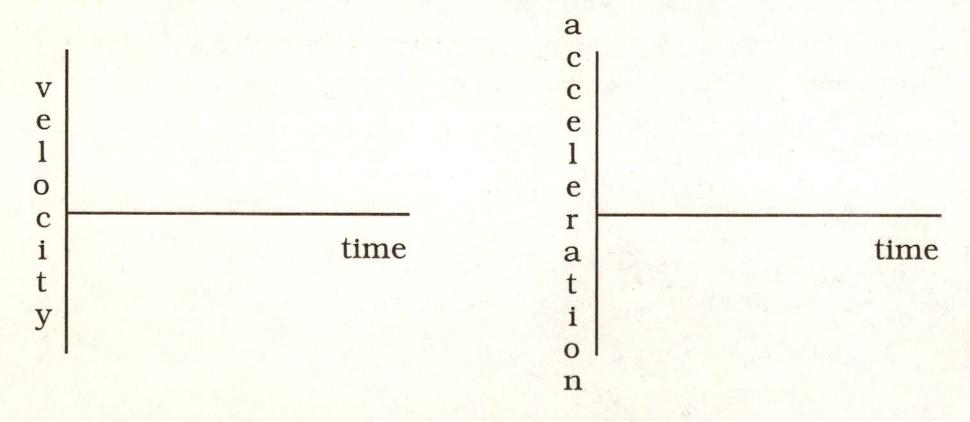

If the acceleration were doubled, how would the graphs be different?

Click **Reset**. Keep the same value of the initial velocity of the car and set the acceleration to 0.60 m/s^2. Click **Run** to test your predictions.

Have the acceleration and velocity graphs changed in the way that you thought they would?

Predict what the velocity and acceleration graphs will look like if the initial velocity is 5 m/s and the acceleration is -1.0 m/s^2 and sketch them in the area below.

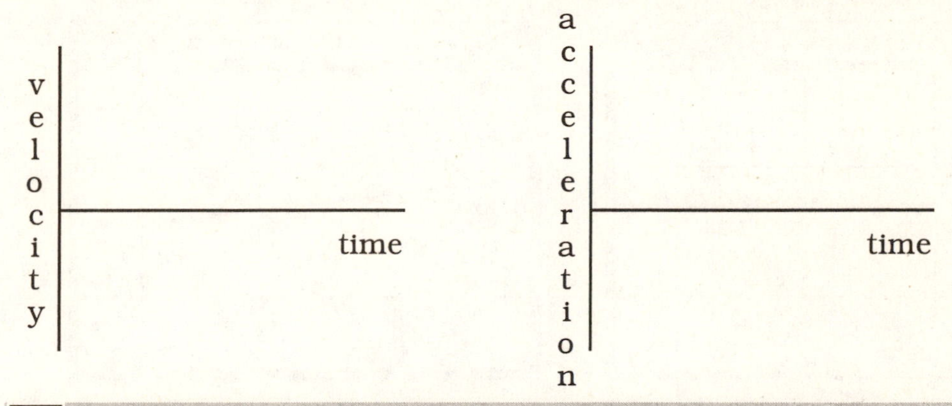

Click **Reset**. Set the initial velocity of the car to 5 m/s and the acceleration to -1.0 m/s^2. Click **Run** and observe the motion of the car until the simulation stops.

Sketch the graphs from the simulation data in the area below and then answer the questions on the following page.

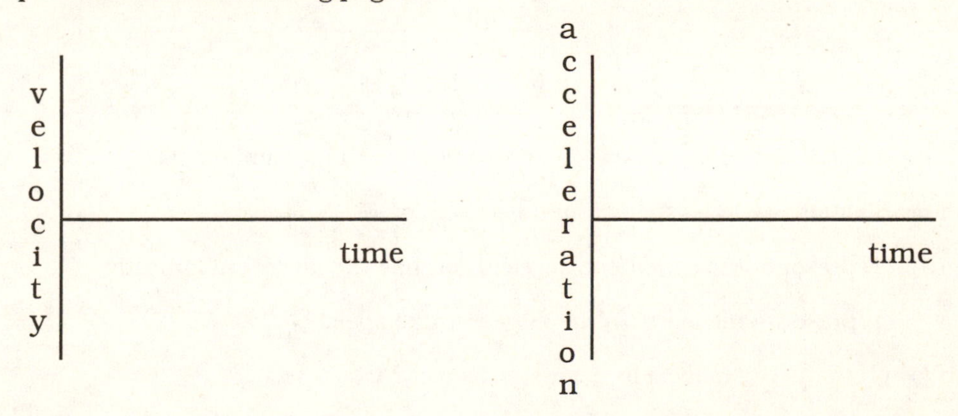

What differences (if any) are there between your predictions and the graphs from the simulation?

Mark each graph with an X (on the time line) at the point where the car turns around. *Does the velocity change at that point? What happens to the acceleration at that point?*

Self-Test Questions for Simulation 2

All the questions refer to these graphs of velocity vs. time for a car moving in one dimension.

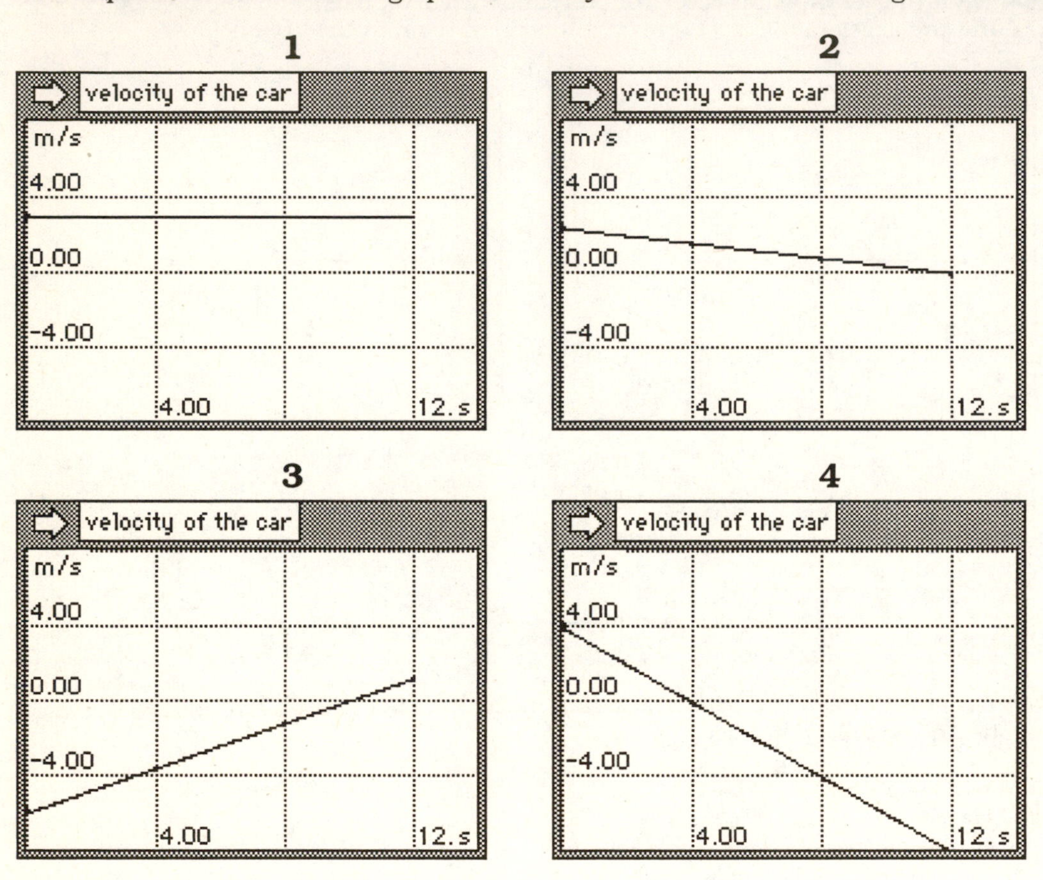

1. In which graph(s), if any, does the car come to a stop (even momentarily)?
2. In which graph(s), if any, does the car change directions?
3. Which graph represents the car whose acceleration has the largest magnitude?
4. Which graph represents the car with the largest initial speed?
5. In which graph is the car moving at a constant speed?

SIMULATION 3 A CAR MOVING WITH CONSTANT ACCELERATION (GRAPHS)

Physics Review

When an object is confined to move in one dimension with a constant acceleration, its velocity and position can be determined as a function of time. Again, we will use the scalar forms of the equations relating position, velocity, and acceleration.

Formulas that you should be familiar with:

$v = v_o + at$ $x = x_o + v_o t + \frac{1}{2}at^2$

Reference topics: one-dimensional motion, constant acceleration

Simulation Details

In the prior two simulations you investigated the relationships between the velocity and acceleration of a car moving in one dimension. Now you will study position too. The initial position of the car is $x = 0.0$ m, and the simulation will automatically stop when 10 seconds have elapsed.

Set the initial velocity to 5.0 m/s and the acceleration to zero. Do not run yet.

Predict what the graphs of velocity and position will look like and draw them in the space below.

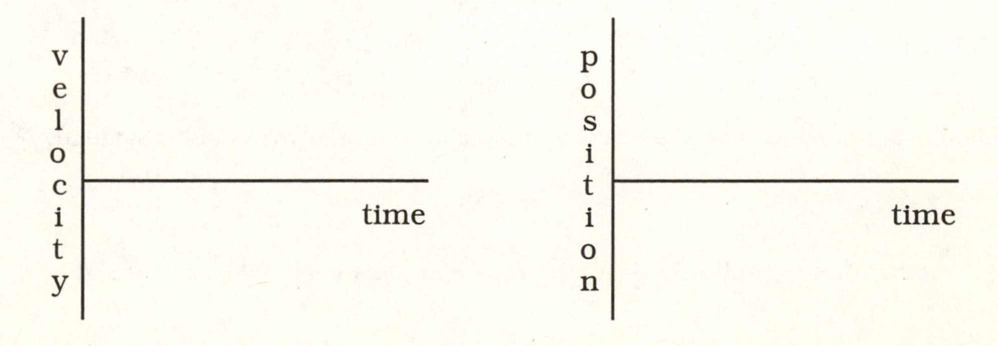

Click Run and observe the motion until the simulation stops.

What differences (if any) are there between your predictions and the graphs in the simulation?

Look at the graphs in the simulation. *Is either graph a straight line? Why or why not?*

Click Run. Set the initial velocity to 2.0 m/s and the acceleration to 1.0 m/s^2. Do not run yet.

Predict what the graphs of velocity and position will look like and draw them in the space below.

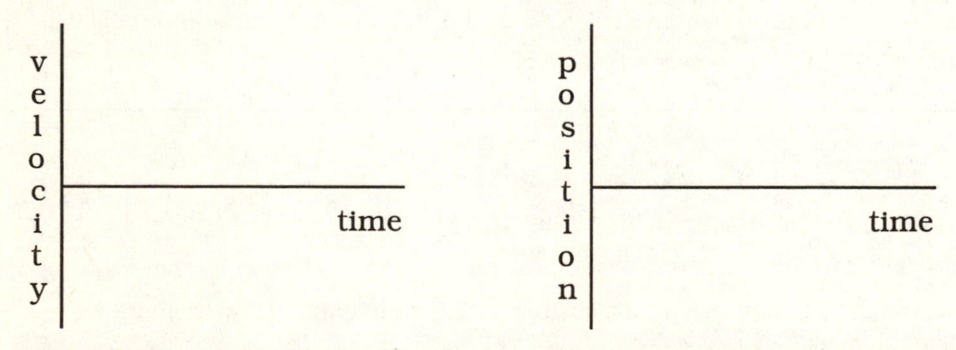

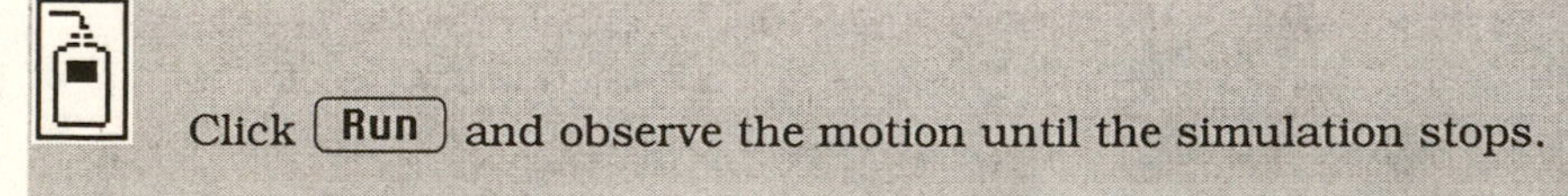

Click Run and observe the motion until the simulation stops.

What differences (if any) are there between your predictions and the graphs in the simulation?

Look at the graphs in the simulation. *Is either graph a straight line? Why or why not?*

Assume an initial velocity of 3 m/s and calculate the acceleration needed to have the car truly stop (i.e., $v = 0$) when the simulation stops ($t = 10$ s).

acceleration = ________

Calculate how far it will have traveled in that case.

distance traveled = ________________

Click **Reset**. Set the initial velocity to 3 m/s and the acceleration to the value that you just calculated. Click **Run** and observe the motion until the simulation stops.

From the simulation data, record or calculate the following:

final velocity of the car = ___________

distance traveled = ___________

Click **Reset**. Set the initial velocity to –5 m/s and the acceleration to 1.0 m/s^2. Do not run yet.

Predict what the graphs of velocity and position will look like and sketch below.

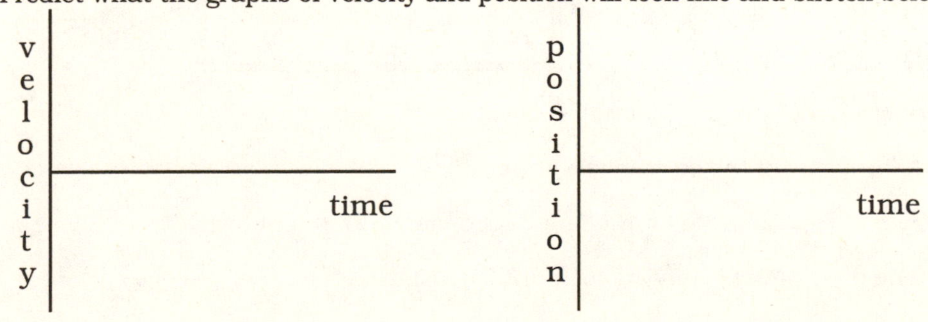

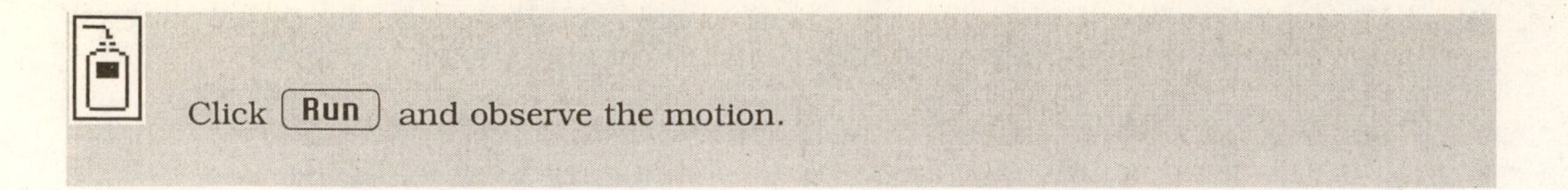

Sketch the simulation graphs in the space below. Mark each graph with an X where the car turns around. Mark the position graph with a Y where the car is back at its initial position.

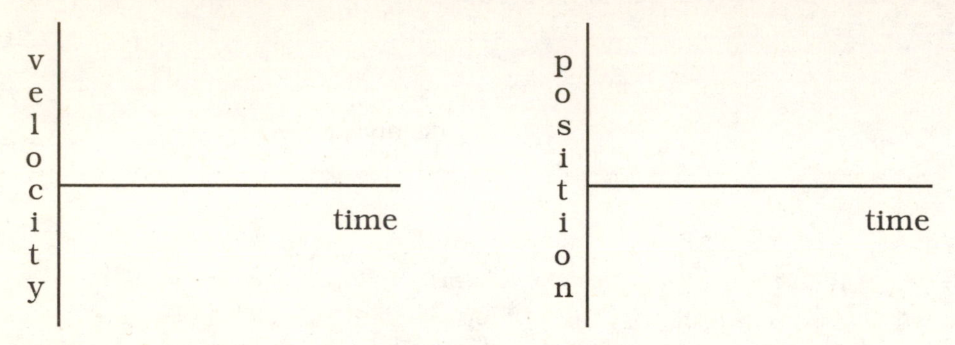

Crossing the axis means something different for position and velocity graphs. *What does it mean in each case?*

Self-Test Questions for Simulation 3

All questions refer to these graphs of the position and velocity of a car.

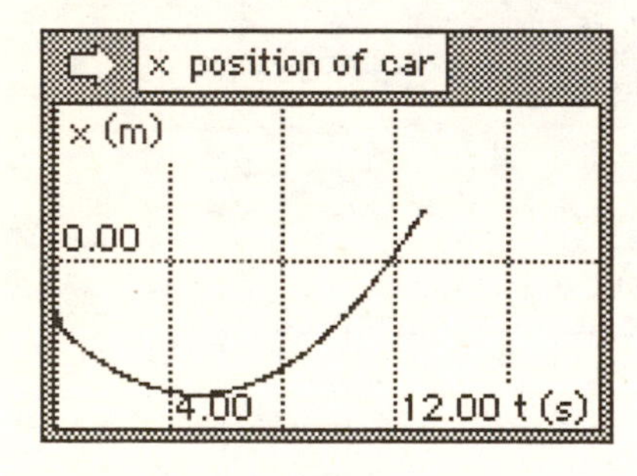

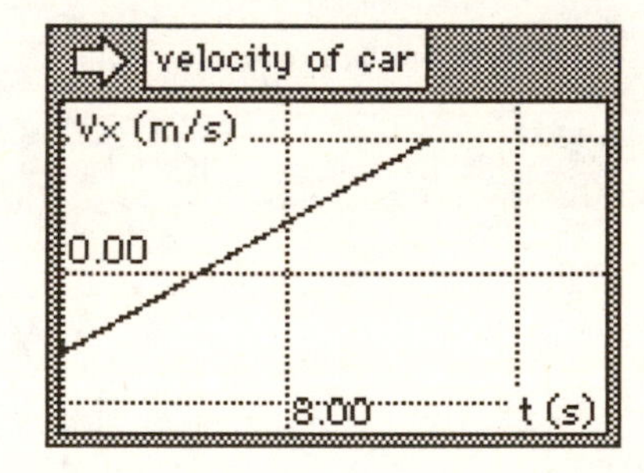

1. Is the acceleration of the car positive or negative?
2. Is the initial velocity of the car positive or negative?
3. Which way is the car moving for the first 5 seconds—to the right or left?
4. At what time does the car come momentarily to a stop?
5. Does the car ever return to its initial starting position? If so, at what time?

SIMULATION 4 THREE CARS RACING

Physics Review

In the previous simulations you set up situations for cars moving in one dimension. You decided on the initial velocity and the acceleration and then you predicted and studied the resulting motion. You looked at numbers and vectors and graphs of position and velocity. Now you are going to go the other way and figure out the accelerations and initial velocities given only the position data of three cars. You can approximate the velocity at a certain time, by using the first formula shown below. For example, if the position is 0.1 meters at $t = 0.20$ seconds and 0.4 meters at $t = 0.30$ seconds, then

$$v = \frac{\Delta x}{\Delta t} = \frac{x_2 - x_1}{t_2 - t_1} = \frac{0.4 - 0.1}{0.3 - 0.2} = 3$$

gives the approximate velocity at $t = 0.25$ seconds (the midpoint of the time interval). Once you have calculated the velocity at different times, you can use the second equation, shown below, to calculate the acceleration in the same manner.

Formulas you should be familiar with:

$$\vec{v} = \frac{\Delta \vec{x}}{\Delta t} \qquad \vec{a} = \frac{\Delta \vec{v}}{\Delta t} \qquad v = v_o + at \qquad x = x_o + v_o + \tfrac{1}{2}at^2$$

Reference topics: one-dimensional motion, constant acceleration

Simulation Details

In this simulation three cars will be racing. You are not given any information on the initial velocities or whether or not the cars are capable of accelerating (they might have broken accelerator pedals). You must watch them move across the computer screen and try to figure out what each one is doing. Tracking will be on, therefore you will have snapshots (an on-screen record) of each car at regular time intervals. The simulation will pause after 1 second has elapsed and then again after 3.2 seconds have elapsed. To go on from either pause, just click on the Run button. The simulation will stop when one of the cars goes off the screen. You will probably need to use the tape player often. The [tape icon] icon will remind you.

Click **Run** to begin watching the car race. The simulation will pause when 1 second has elapsed. Before going on, answer the following questions.

Which car has traveled the greatest distance at this time?

Can you tell which car is accelerating?

Click **Run** to continue watching the car race. The simulation will pause again at 3.2 seconds. Before going on, answer the following questions.

Can you tell now which car is accelerating?

Which car has gone the farthest?

Do you think that the red car will win the race?

Click **Run** to continue watching the car race to the end. Look at the tracks of the cars and answer the following questions.

Which cars are not accelerating?

What is the approximate ratio of their velocities?

From the simulation data for the yellow car, make a plot of its position vs. time.

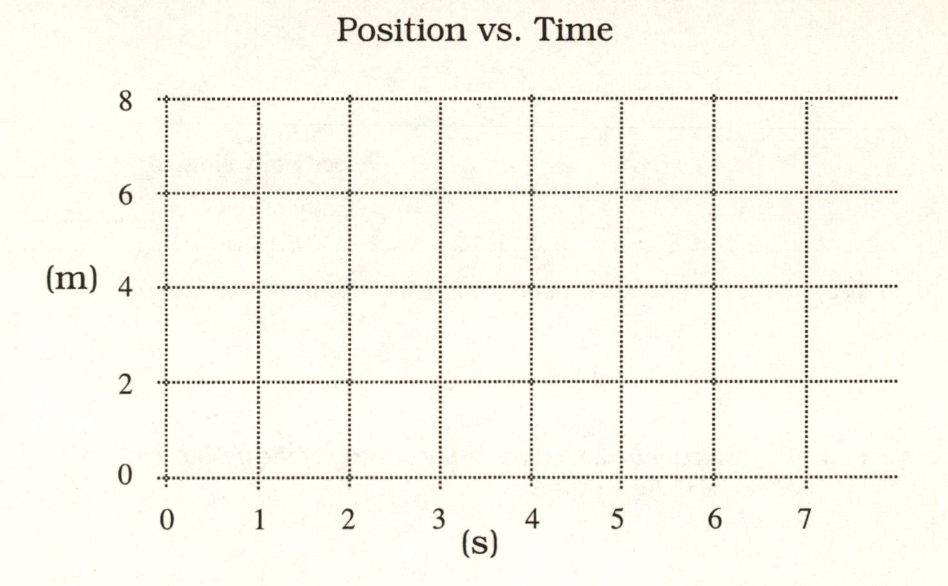

From the simulation data for the red car, make a plot of its position vs. time.

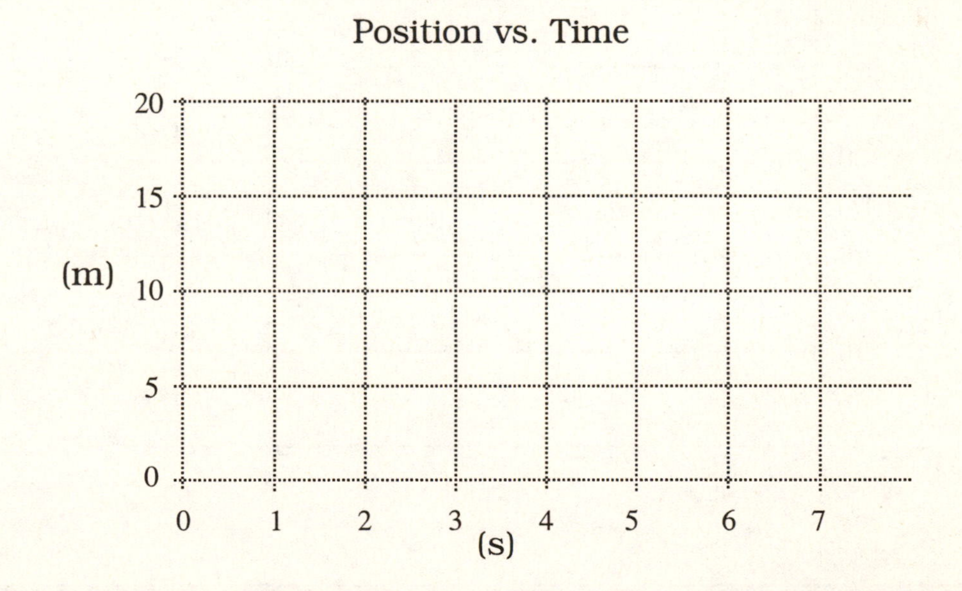

From your position data and the graphs on the previous page, determine the velocity of the yellow car and the red car.

velocity of yellow car = ____________

velocity of red car = ______________

Did you predict that these cars had constant velocities? How did you know?

What is the ratio of the velocity of the red car to the velocity of the yellow car? Is this what you determined by looking at the tracks?

From the simulation data for the green car, make a plot of its position vs. time. Use five or six data points.

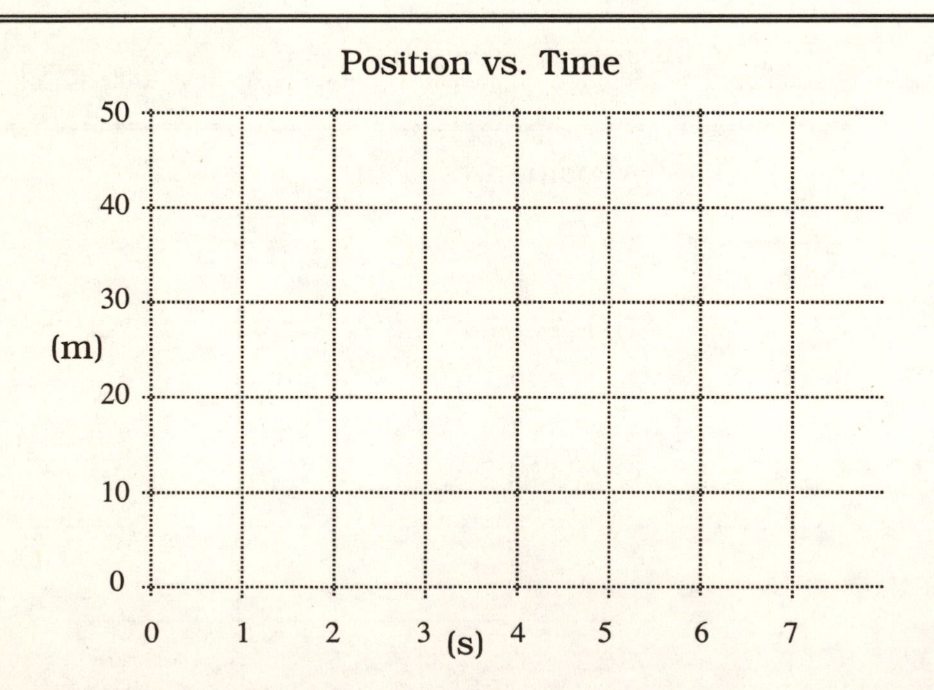

From the position data, calculate the approximate velocity of the green car at the following times. Use the method described in the physics review section on page 13. Remember, the smaller the value of Δt, the more accurate your velocity value will be.

t = 1.5 s

$v(1.5)$ = __________

t = 3.5 s

$v(3.5)$ = __________

t = 5.0 s

$v(5.0)$ = __________

t = 6.5 s

$v(6.5)$ = __________

Plot the values that you just calculated on the graph below.

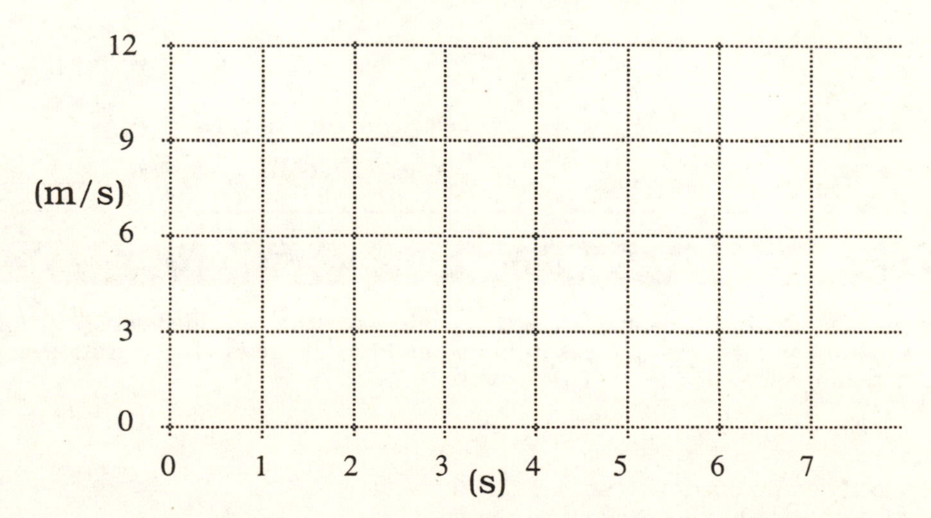

Is the green car accelerating? How do you know?

What is the approximate acceleration of the green car?

What is the initial velocity of the green car?

Use the [Q_O] to find the point where the green car passed the red car. Record the time and the position of each car.

time = _______

red car position = __________

green car position = ________

From this data, determine the velocity of the red car and the acceleration of the green car. Assume that the initial velocity of the green car is zero.

velocity of the red car = ___________________

acceleration of the green car = ____________

Self-Test Questions for Simulation 4

Three cars all start from the same starting line and move in one dimension. Car A has a constant speed of 4 m/s, car B starts from rest and has a constant acceleration, and car C has a constant speed of 3 m/s. True or false?

1. The velocity graph of car A is a horizontal line.
2. The velocity graph of car B is not a straight line.
3. Car B can never catch up with car A.
4. Car C can never catch up with car A.
5. During any time interval, car B goes the farthest.

SIMULATION 5 A BOAT CROSSING A RIVER

Physics Review

Velocity is a vector quantity. It is defined by a magnitude ($|V|$) and a direction (θ). Adding (or subtracting) vector quantities is not the same as adding (or subtracting) numbers, and there are several ways to do it. One way is to "break each vector into its components," add the components separately, and then put those new components back together as a vector. This is illustrated below for $\vec{C} = \vec{A} + \vec{B}$.

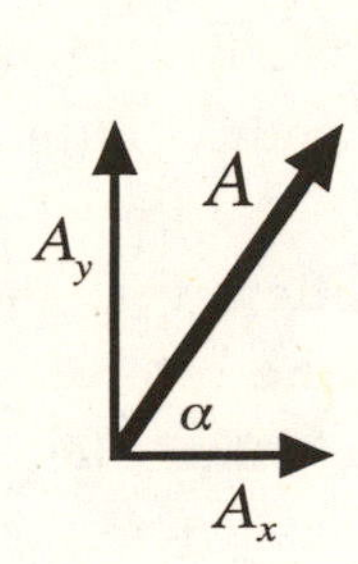

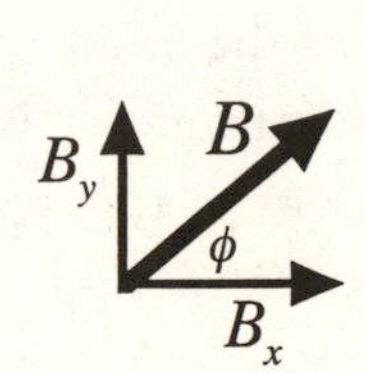

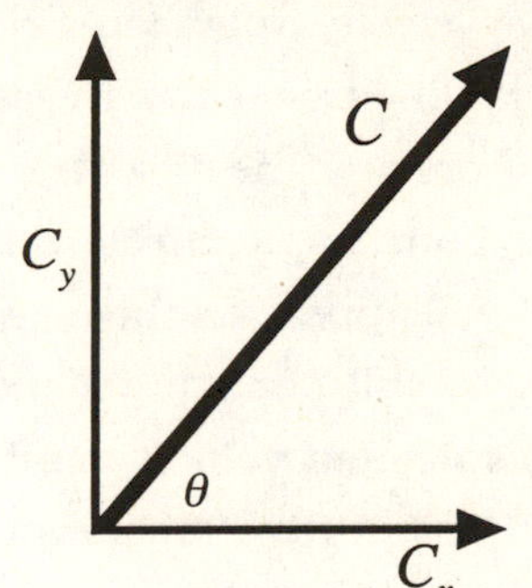

$$A_x = |A|\cos\alpha$$
$$A_y = |A|\sin\alpha$$

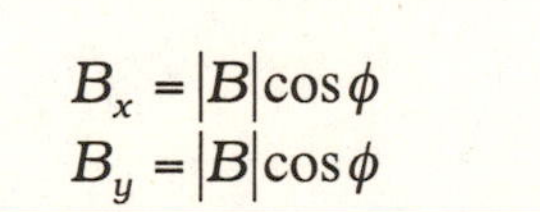

$$C_x = |C|\cos\theta$$
$$C_y = |C|\sin\theta$$

$$C_x = A_x + B_x \qquad C_y = A_y + B_y$$

$$|C| = \sqrt{C_x^2 + C_y^2} \qquad \tan\theta = \frac{C_y}{C_x}$$

A boat traveling across a river is an example of a problem where velocity vectors must be added to accurately describe the motion of the boat. We will denote the velocity that the <u>boat</u> can travel in still <u>water</u> as $\vec{v}_{bw}$ and the velocity of the river (<u>water</u> with respect to the <u>shore</u>) as $\vec{v}_{ws}$. Then the velocity of the <u>boat</u> with respect to the <u>shore</u> is given by $\vec{v}_{bs} = \vec{v}_{bw} + \vec{v}_{ws}$. Using the graphical way of adding vectors, you place the tail of one on the head of the other and draw the resulting sum.

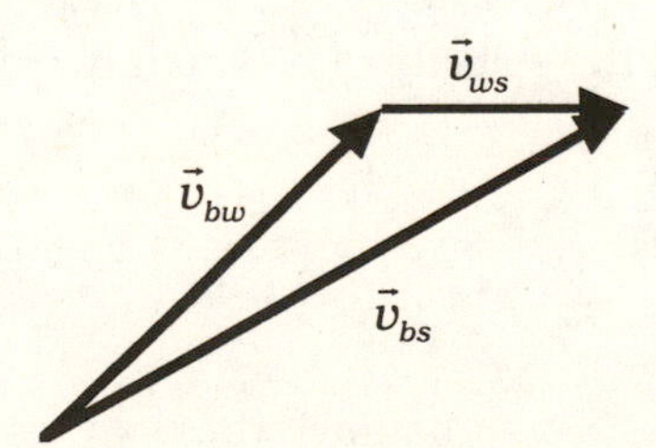

Reference topics: vectors, vector addition, relative velocity

Simulation Details

A boat is set to cross a river 100 meters wide. The positive y direction will be taken as north. The boat's speed in still water is set to 20 m/s, but its direction is adjustable. For this simulation we will define the boat's heading θ as the angle the velocity vector makes with the y axis. Positive angles will represent boats moving to the northeast (upward and to the right on your screen) and negative angles will represent boats moving to the northwest (upward and to the left on your screen). A positive velocity for the river will indicate that the river is flowing to the east and a negative velocity will mean a river flowing to the west. The path of the center of the boat will be tracked as the simulation progresses. You can view the motion of the boat from any one of several defined reference frames by clicking on one of the buttons. Your frame of reference will not change until you click another button. The "eye" will appear on the object selected as the frame of reference. The trace left by the boat will change depending on your frame of reference. **However, the vector displays of velocity and all numerical outputs will be in the reference frame of the person (observer) waiting to be picked up on the other shore.** This simulation will automatically stop when the boat has reached the other shore.

Set the river current speed to zero. Set the boat's heading to 0°. Click **Run** and observe the motion until the boat gets across the river.

From the simulation data, record or calculate the following:
(wrt = with respect to)

time to cross the river = ____________

x component of the velocity of the boat (wrt the shore) = _________

y component of the velocity of the boat (wrt the shore) = _________

$|\vec{v}_{bw}|$ = _______

$|\vec{v}_{bs}|$ = _____________

total distance the boat traveled = ____________

Click **Reset**. Set the river current speed to 8 m/s (to the east) and leave the boat's heading set to 0°. Click **Run**.

From the simulation data, record or calculate the following:

time to cross the river = ____________

x component of the velocity of the boat (wrt the shore) = _________

y component of the velocity of the boat (wrt the shore) = _________

$|\vec{v}_{bw}|$ = ______________

$|\vec{v}_{bs}|$ = ______________

total distance the boat traveled = _____________

The boat landed ___________ meters to the (*east, west*) of the person at the shore.

For the vectors shown below, label each one ($\vec{v}_{bw}$, $\vec{v}_{ws}$, $\vec{v}_{bs}$) and give numerical values for its magnitude and each component.

= +

NAME	______	______	______
magnitude	______	______	______
x component	______	______	______
y component	______	______	______

Describe or sketch the path of the boat with respect to the person at the shore.

Click **Reset**. Change the reference frame to the floating fishing boat. Its speed in the water is zero. Click **Run**.

Did any of the following change when you changed the reference frame? Why or why not?

time to cross the river:

x component of the velocity of the boat (wrt the shore):

y component of the velocity of the boat (wrt to the shore):

$|\vec{v}_{bw}|$:

$|\vec{v}_{bs}|$:

total distance the boat traveled:

how far east or west of the observer at the shore the boat landed:

Describe or sketch the path of the boat with respect to the floating fishing boat.

Click **Reset**. Click **Jogger** to change to the reference frame of the jogger. Her speed is 3 m/s to the west. Click **Run**.

Describe or sketch the path of the boat with respect to the jogger.

Assume a boat speed of 20 m/s and a river current speed of 8 m/s (to the east). Calculate the boat's heading needed to insure a landing near the person waiting to be picked up across the river on the other shore. You will need accuracy of two digits after the decimal point.

θ = __________

Click **Reset**. Click on **Observer** to set the reference frame to the person at the shore. Set the river current speed to 8 m/s and the boat's heading to the value that you just calculated (type it in the box). Click **Run**.

Did the boat reach the person at the shore? If not, redo your calculation and try it again.

From the simulation data, record or calculate the following:

time to cross the river = ____________

x component of the velocity of the boat (wrt the shore) = _________

y component of the velocity of the boat (wrt the shore) = _________

$|\vec{v}_{bw}|$ = ______

$|\vec{v}_{bs}|$ = ____________

total distance the boat traveled = ____________

The boat landed ___________ meters to the (*east, west*) of the person at the shore.

Compare these values to those that you recorded on pages 20 and 21. Only one thing has remained constant. *What is it?*

For the vectors shown below, label each one ($\vec{v}_{bw}$, $\vec{v}_{ws}$, $\vec{v}_{bs}$) and give numerical values for its magnitude and each component.

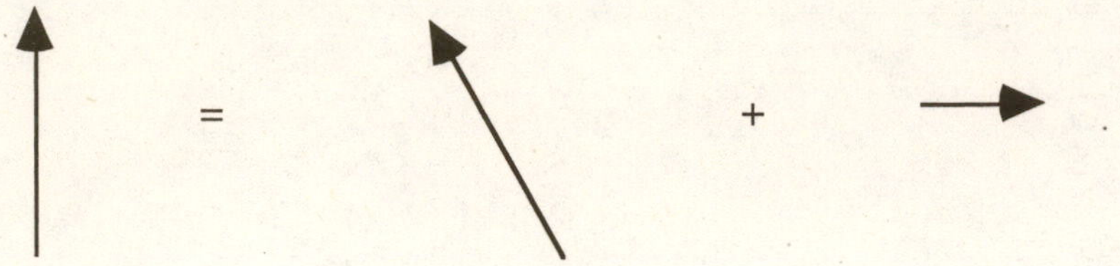

= +

NAME	______	______	______
magnitude	______	______	______
x component	______	______	______
y component	______	______	______

Hint

If you are having trouble with this part, try one of the following:

To "see" the vector and the corresponding numerical values of the velocity of the boat with respect to the water, reset and set the river velocity to zero.

Work backward from the fact that the sum of the x components of $\vec{v}_{bw}$ and $\vec{v}_{ws}$ must be equal to the x component $\vec{v}_{bs}$. Ditto for the y components.

Self-Test Questions for Simulation 5

The following statements apply to a boat traveling across a river. The speed of the boat in still water is 20 m/s, and the river is flowing with a speed of 5 m/s. True or false?

1. The velocity of the boat with respect to the shore depends on the velocity of the river.
2. It is not possible for the boat to arrive at a point directly across the river.
3. The motion of the boat is different for different observers.
4. The time that it takes the boat to cross the river is the same regardless of who is measuring it (e.g., the person in the boat or the person at the shore).
5. The total distance that the boat travels in crossing the river depends on the speed of the river and the heading of the boat.

SIMULATION 6 AN AIRPLANE FLYING WITH WIND

Physics Review

An airplane flies through the air with a velocity that we will denote by $\vec{v}_{pa}$ (velocity of the plane with respect to still air). During many flights, however, the air is moving (wind) and the pilot must be able to steer the plane in a direction that will allow it to reach a certain point, taking into account this wind. We will denote the velocity of the wind by $\vec{v}_{ag}$ (velocity of the air with respect to the ground). To get $\vec{v}_{pg}$, the velocity of the plane with respect to the ground, you must add the two vector quantities.

$$\vec{v}_{pg} = \vec{v}_{pa} + \vec{v}_{ag}$$

You can either "break each vector into x and y components" and then add the components together or use a graphical "tail to tip" method. See the previous simulation for more details. A sample vector addition is shown graphically, below.

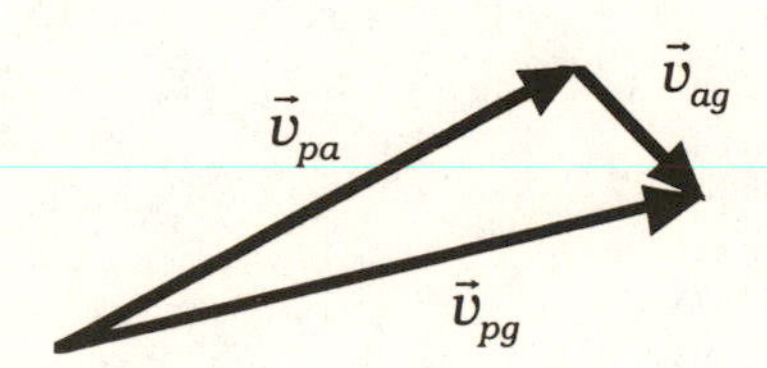

Reference topics: vectors, vector addition, vector subtraction

Simulation Details

In this simulation an airplane is trying to land at a nearby airport. The speed of the airplane is fixed at 206 m/s. A jet can attain such airspeed, but it is rather unrealistic for a plane that is about to land in a few seconds. In order to visualize the vectors and see the airport on the screen, and not take too long for the plane to "land," this high speed is used in this simulation. You can adjust the aircraft heading by typing an angle into the input box labeled Aircraft Heading. The angle must have a value between 0° and 360° and it is measured with respect to the positive x direction. An angle of 90° will represent a plane heading north and an angle of 315° will represent a plane heading southeast. On your screen you will see the velocity vector of the airplane and the numerical values of its magnitude and components. These values are with respect to the ground. You will also see a vector representing the velocity of the air (magnitude and direction of the wind). This is also with respect to the ground. If you want to see the velocity vector (and the corresponding numerical values) of the plane in still air, set the magnitude of the wind to zero. The initial position of the airplane is (0,0) and the airport is located 400 meters to the east of there and 100 meters to the north (400,100). This simulation will stop when the airplane has traveled a distance of about 400 meters in the x direction. An output meter will also give the x and y position of the airplane and the distance traveled.

How far away is the airport?

Assuming no wind, calculate how long it will take the plane to get to the airport.

time = ________

Assuming no wind, calculate in which direction the plane must head in order to land at the airport.

θ = __________

Set the aircraft heading to the value that you just calculated (type it into the box) and click **Run**.

The simulation will stop when the horizontal distance traveled is just greater than 400 m. Use the [O O] to go back one frame if that gives you an x position that is closer to 400 m.

From the simulation data, record or calculate the following:

the time = ________________

the x position of the airplane = ________________

the y position of the airplane = ________________

For our purposes we will consider that the plane landed at the airport if it is within 10 meters. *Did the plane land at the airport?*

Does the simulation value for the time to get to the airport agree with your calculation on the previous page?

Now we examine what will happen to the airplane if there is a 50 m/s wind blowing. First we will assume that the pilot of the plane knows nothing about the wind: then we will see how far off course it carries him and in what direction. Let's assume that the wind is coming from the west. *Once the airplane has traveled 400 meters to the east, do you think that the plane will be to the north of the airport or to the south of the airport?* (Note: Do not run the simulation to answer this question. Try to reason it out yourself.)

Click **Reset**. Leave the aircraft heading set to the same value as before (when there was no wind). Set the wind speed to 50 m/s and the direction of the wind to 0°. Click **Run** and observe the motion.

Use the [rewind button] to go back one frame if that gives you an x position that is closer to 400 m.

From the simulation data, record or calculate the following:

the time = ______________

the x position of the airplane = ______________

the y position of the airplane = ______________

Is the plane south or north of the airport?

About how far from the airport is it?

Is this what you expected would happen?

In the space below, draw the vectors representing the following quantities.

$$\vec{v}_{pg} = \vec{v}_{pa} + \vec{v}_{ag}$$

Fill in the corresponding numerical values.

	$\vec{v}_{pg}$	$\vec{v}_{pa}$	$\vec{v}_{ag}$
x component	______	______	______
y component	______	______	______
magnitude	______	______	______
θ	______	______	______

Click [Reset]. Leave the aircraft heading set to the same value as before (when there was no wind). Set the wind speed to 50 m/s and the direction of the wind to 180°. Now the wind is coming from the east. Click [Run] and observe the motion.

Use the [QO] to go back one frame if that frame gives you an *x* position that is closer to 400 m. From the simulation data, record or calculate the following:

the time = ____________

the *x* position of the airplane = ____________

the *y* position of the airplane = ____________

Is the plane south or north of the airport?

About how far from the airport is it?

Is this what you expected would happen?

How far did the plane travel?

Calculate its speed from the value of distance traveled divided by time.

speed = ____________

In the space below, draw the vectors representing the following quantities.

$$\vec{v}_{pg} \quad = \quad \vec{v}_{pa} \quad + \quad \vec{v}_{ag}$$

Fill in the corresponding numerical values.

x component	______	______	______
y component	______	______	______
magnitude	______	______	______
θ	______	______	______

Circle the value above that is the closest to the speed that you calculated at the bottom of the previous page. That is the speed of the plane with respect to the ground.

Now we will look at what the pilot has to do if he knows about this wind and wants to compensate for it. Let's go back to the case where the wind is blowing from the west. Remember that we are trying to find the aircraft heading necessary to land the plane at the airport. Then in vector form (with components shown):

$$\vec{v}_{pg} \quad = \quad \vec{v}_{pa} \quad + \quad \vec{v}_{ag}$$

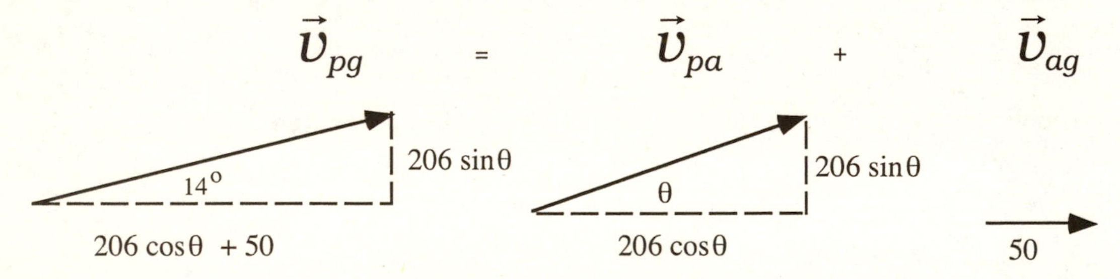

The equation to solve for the angle θ is:

$$\frac{206 \sin\theta}{206 \cos\theta + 50} = \tan(14°) = 0.25$$

You can solve this analytically (but it is tedious). You can use a calculator or computer, or guess a sensible solution and show that it works.

Click **Reset**. Set the aircraft heading to the value that you think is correct. Set the wind speed to 50 m/s and its direction to 0°. Click **Run**.

Use the [step back button] to go back one frame if that frame gives you an x position that is closer to 400 m. From the simulation data, record or calculate the following:

the time = ________________

the x position of the airplane = ________________

the y position of the airplane = ________________

distance that the plane traveled = ________________

Did the plane land at the airport?

Calculate its speed from the distance traveled divided by the time.

speed = ____________

	$\vec{v}_{pg}$	=	$\vec{v}_{pa}$	+	$\vec{v}_{ag}$
x component	______		______		______
y component	______		______		______
magnitude	______		______		______
θ	______		______		______

Circle the speed that is the closest to the speed that you calculated above. That is the speed of the plane with respect to the ground.

What about north and south winds? Use the simulation to answer the following questions.

If the wind is 50 m/s blowing from the south and the plane heads straight for the airport, how far off course is it when the simulation stops?

If the wind is 50 m/s blowing from the south, in what direction must the plane head to land at the airport?

If the wind is 50 m/s blowing from the north and the plane heads straight for the airport, how far off course is it when the simulation stops?

If the wind is 50 m/s blowing from the north, in what direction must the plane head to land at the airport?

Now we will look at the case where the wind is coming from the northwest.

Click **Reset**. Set the wind speed to 50 m/s and its direction to 330°. Adjust the direction of the aircraft heading until you think that the plane will land at the airport. Record this angle below and then click **Run**.

Use the [button] to go back one frame if that frame gives you an *x* position that is closer to 400 m. From the simulation data, record or calculate the following:

aircraft heading = ______________

the time = ______________

the *x* position of the airplane = ______________

the *y* position of the airplane = ______________

Did the plane land at the airport?

In the space below, draw the vectors representing the following quantities.

$$\vec{v}_{pg} = \vec{v}_{pa} + \vec{v}_{ag}$$

Fill in the corresponding numerical values.

x component	______	______	______
y component	______	______	______
magnitude	______	______	______
θ	______	______	______

Do the values in the last two columns add up to the value in the first column for

- the x component of velocity?
- the y component of velocity?
- the magnitude of the velocity?
- the angle θ?

Look back at the data you recorded in the same table as above on previous pages (28, 30, 31). Only two things are the same on all of the pages. *What are they?*

Self-Test Questions for Simulation 6

The questions refer to the three vectors, A, B, and C. One vector represents the velocity of the plane in still air, one represents the velocity of the wind, and one represents the velocity of the plane with respect to the ground. The airplane is traveling from a point O and arrives at an airport that is 1000 meters away (straight line). The air speed of the plane is 206 m/s. All numbers have units of m/s.

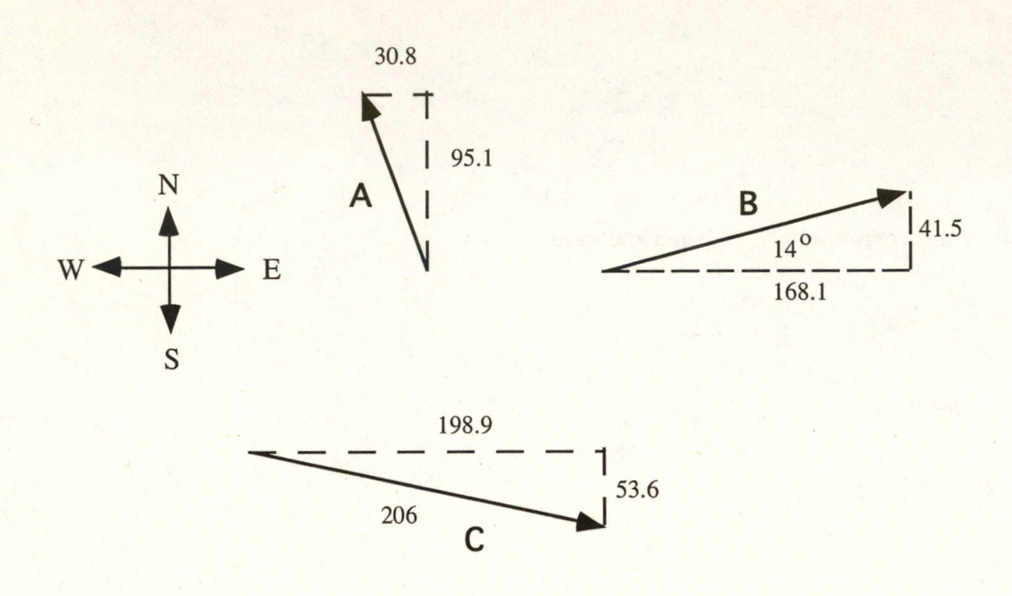

1. Which is correct, $\vec{C} = \vec{A} + \vec{B}$ or $\vec{B} = \vec{A} + \vec{C}$?
2. What is the speed of the wind?
3. What is the speed of the plane with respect to the ground?
4. How long does it take to get from point O to the airport?
5. What direction is the wind coming from?

SIMULATION 7 A BALL THROWN STRAIGHT UP (VECTORS)

Physics Review

A special case of motion with constant acceleration is that of a ball thrown straight up. The ball will rise to some maximum height in some time and then fall back down to the ground. In the idealized case the only acceleration that acts to change the velocity of the ball is that of gravity, indicated by g. The acceleration due to gravity on Earth has a value of approximately 9.8 m/s^2 (*Interactive Physics* uses 9.81 m/s^2) and on the Moon a value of about 1/6 of that.

Formulas that you should be familiar with:

$$v = v_o - gt \qquad y = y_o + v_o t - \tfrac{1}{2} g t^2 \qquad v^2 = v_o^2 - 2g(y - y_o)$$

By writing these formulas this way (and in scalar form), we take the positive direction to be upward. That is, the initial velocity of the ball will be positive and the acceleration due to gravity whether on Earth or the Moon will be a positive number when inserted into either of these formulas.

Reference topics: free fall, falling bodies

Simulation Details

You will be able to adjust the speed with which the ball is thrown up, and you will be able to switch back and forth between the Earth and the Moon using the "Gravity Switch" slider. The simulation will automatically pause when the ball gets to the top of its flight. To continue from there, click the Run button. To run from the beginning, click Reset and then click Run. The initial position of the ball is y = 0.0 m. The simulation will stop when the ball is back near its original position.

Make sure that the gravity switch is set to Earth. Set the ball's initial speed to 10 m/s. Do not run yet.

Assume an initial speed of 10 m/s and calculate the velocity at t = 0.4 seconds.

velocity = ___________

Click Run and observe the motion for about 0.5 seconds. Then use the tape player to go back to the frame where the time is 0.4 seconds.

From the simulation data, record the velocity and the height of the ball at that point.

velocity = ___________

height = ___________

How does your calculation of velocity compare with what you just recorded from the simulation data?

Calculate how high the ball will go.

maximum height of the ball = ___________

Calculate how long it will take for the ball to get to that height.

time to the top = ___________

Click [Reset] then click [Run] and observe the motion until the simulation pauses.

From the simulation data, record the time and maximum height, and compare to your calculations on the previous page.

time = _______

maximum height of the ball = ____________

Click [Reset]. This time when you run the simulation, observe the velocity and acceleration vectors on the way up. Click [Run].

Note that the velocity is positive (upward is defined as positive) and changing, while the acceleration is negative and constant. Draw the velocity and acceleration vectors at the times given in the space below:

when the ball is at the initial position:

when the ball is halfway up:

when the ball is at the top:

Make a general statement about the velocity of an object that applies whenever its velocity and (constant) acceleration vectors are opposite in direction.

Calculate how long it will take for the ball to fall back to the starting point.

time from the top to the starting point = __________

Click **Run** to see the ball fall back to its starting point.

How does the time that you just calculated compare with the time to fall shown in the simulation?

Click **Reset**. This time when you run the simulation, observe the velocity and acceleration vectors on the way down. Click **Run**.

Note that the velocity is negative (upward is defined as positive) and changing and the acceleration is negative yet constant. Draw the velocity and acceleration vectors at the times given in the space below:

when the ball is at the top:

when the ball is halfway down:

when the ball is back in his hand:

Make a general statement about the velocity of an object that applies whenever its velocity and (constant) acceleration vectors are in the same direction.

Use the [tape player icon] to locate the point where the ball is on the way down and is at a height of about 3.2 m. From the simulation data, record the following:

velocity of the ball = ______________

height of the ball = ______________

Compare this value of velocity to that which you recorded for this same height when the ball was moving up (page 36). *What is the significance of the negative value?*

Click **Reset**. Slide the gravity switch to Moon. Click **Run**. The ball will most likely go off the screen rather quickly. Never mind - *Interactive Physics* is still keeping track of it, and you can see the values of its position and velocity in the meters on the screen. It will pause at the top so you can record values of position and time. Then click **Run**.

From the simulation, record:

maximum height of the ball = ____________

time to the top = ___________

Did the ball go higher on the Moon? How many times higher? Why?

Did the ball take longer to get to the top? How many times longer? Why?

Click [Reset] and slide the gravity switch to Earth. Set the initial speed to 12 m/s and click [Run].

From the simulation data, record or calculate the following:

maximum height = ________

time to get to that height = ________

Calculate the initial speed needed for the ball to rise to that same height on the Moon.

initial speed = _______

Click [Reset] and slide the gravity switch to Moon. Type the value that you just calculated into the initial speed box. Click [Run].

From the simulation data, record or calculate the following:

time to get to that height = ________

Did you expect that it would take longer (to rise to the same height) on the Moon? Why?

Self-Test Questions for Simulation 7

These statements apply to a ball thrown straight up. True or false?

1. The velocity of the ball is zero at the highest point.
2. When the ball returns to its original position, its velocity has the same magnitude as the initial velocity.
3. For the same initial speed, the time it takes for the ball to reach its maximum height on the Earth is 6 times longer than the time that it takes the ball to reach its maximum height on the Moon.
4. On both the Earth and Moon the ball's acceleration is zero at its maximum height.
5. For the same initial speed, the ball goes 6 times higher on the Moon than it does on the Earth.

SIMULATION 8 A BALL THROWN STRAIGHT UP (GRAPHS)

Physics Review

When a ball is thrown straight up and is affected only by the acceleration due to gravity, the ball will rise to some height and then fall back down to the ground. A common misconception is that the acceleration at the top of the flight is zero. In fact, it is not zero; it has a constant value of about 9.8 m/s^2 on Earth. The velocity at the top of the flight, however, is zero and you should have seen that in the previous problem when you studied the acceleration and velocity vectors. The velocity of the ball is positive on the way up and negative on the way down, but the acceleration is always negative.

Formulas that you should be familiar with:

$$v = v_o - gt \qquad y = y_o + v_o t - \tfrac{1}{2}gt^2 \qquad v^2 = v_o^2 - 2g(y - y_o)$$

By writing these formulas this way, we take the positive direction to be upward. That is, the initial velocity of the ball will be positive and the acceleration due to gravity will be a positive number when inserted into either of these formulas.

Reference topics: free fall, falling bodies

Simulation Details

In this problem a ball will be thrown straight up with an initial speed that you can adjust. The acceleration due to gravity is set to the value for the Earth. You will be studying and predicting the graphs of position and velocity as a function of time. The initial position (height) of the ball is y = 0.0 m. This simulation will automatically stop when the ball is back near its original position.

Set the initial speed to 9 m/s. Do not run yet.

Predict what the position and velocity graphs will look like and draw them in the area below. Place an X at the point on each graph indicating the point where the ball is at the top of its flight. The positive direction is taken as upward.

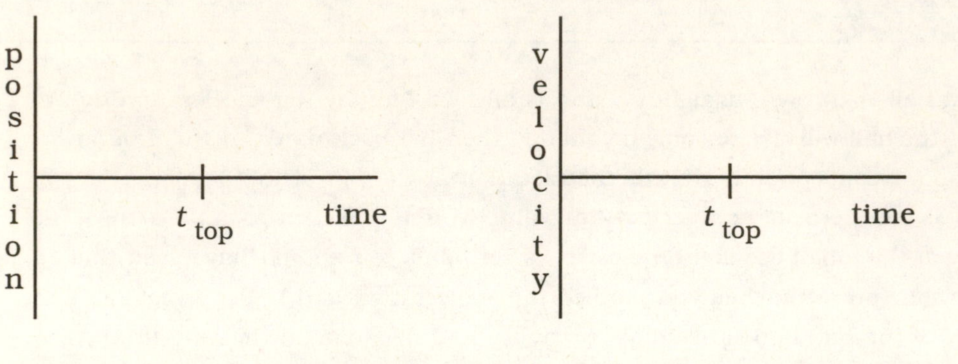

What differences (if any) are there between your predictions and the graphs for position and velocity of the ball shown in the simulation?

What is the value of the slope of the velocity curve?

From the simulation data, record or calculate the following:

time to the top = ____________

velocity at the top = _________

acceleration at the top = ____________

total time = _________

velocity at the return to the starting point = __________

In what follows, circle the correct word. If the initial speed is decreased, then:

- the time to the top will be (*greater than, less than, the same as)* it was in the previous case.
- the velocity at the top will be (*greater than, less than, the same as)* it was in the previous case.
- the acceleration at the top will be (*greater than, less than, the same as)* it was in the previous case.
- the total time for the trip will be (*greater than, less than, the same as)* it was in the previous case.
- the speed at the return to the starting point will be (*greater than, less than, the same as)* it was in the previous case.

Predict what the position and velocity graphs will be for a smaller initial speed and sketch them in the region below. A reference point (t-top) is shown representing the time it took for the ball in the previous case to get to the top.

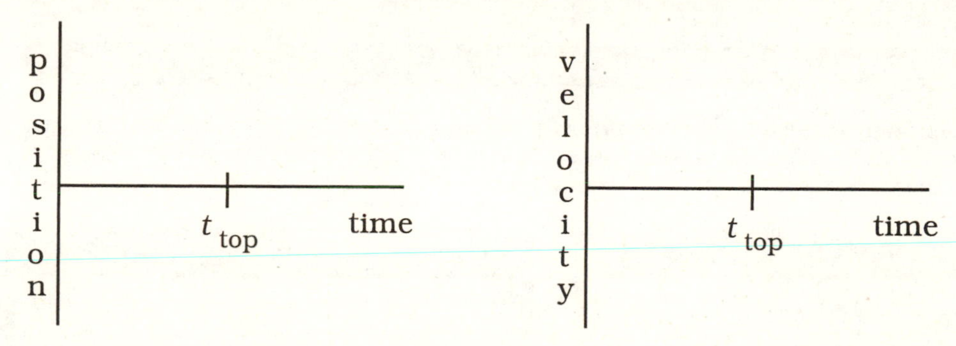

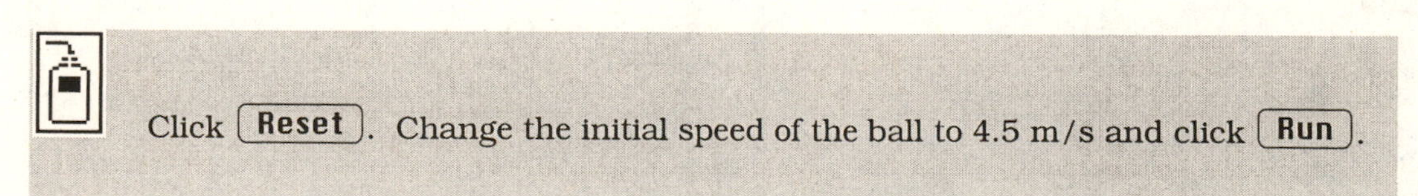

From the simulation data, record or calculate the following:

time to the top = ___________

velocity at the top = ________

acceleration at the top = __________

total time = _______

velocity at the return to the starting point = _________

Compare these values you just recorded to your predictions on the top of the previous page. *What differences (if any) are there between them?*

Compare the graphs shown in the simulation to those that you predicted on the previous page. *What differences (if any) are there between them?*

Optional: Vary the initial velocity of the ball and record the time for the trip and the maximum height attained by the ball. Plot time (vertical axis) vs. initial velocity (horizontal axis), and height (vertical axis) vs. initial velocity (horizontal axis). You can do this by hand, or use a spreadsheet or graphing program if you have one on your computer. To get a good plot, you probably need to have 5 or 6 data points. *Are the graphs what you expected? Are they straight lines? Why or why not?*

Self-Test Questions for Simulation 8

The questions refer to the following graphs of position and velocity for two balls thrown directly up with different initial speeds. Each height curve goes with one of the velocity curves. The initial position of each ball is y = 0.0 m.

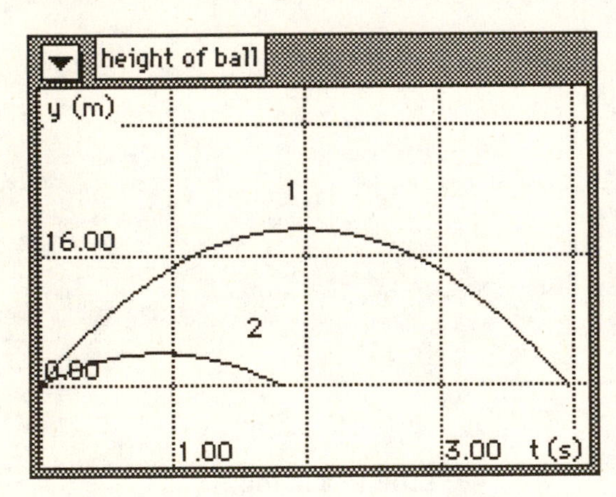

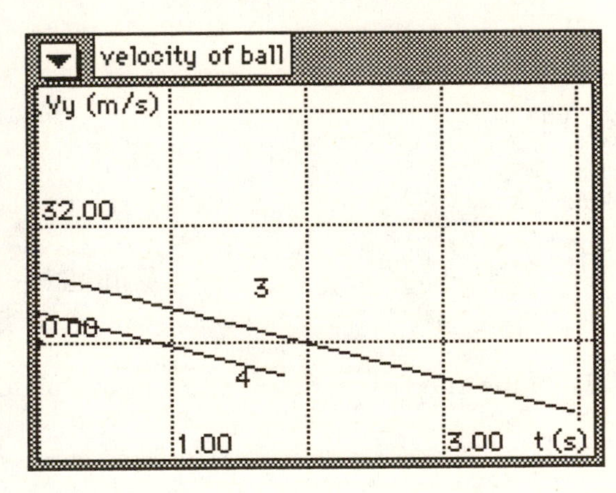

1. Which height curve represents the ball launched with the greatest initial speed?
2. About how long does it take the ball represented by curve 2 to complete its trip and return to the ground?
3. Approximately how high does the ball represented by curve 1 go?
4. What was the approximate initial speed of the ball represented by curve 1?
5. At approximately what other time is the height of the ball represented by curve 1 the same as it was at 1 second?

SIMULATION 9 PROJECTILE MOTION IDEALIZED

Physics Review

Baseballs, basketballs, and golf balls flying through the air are examples of a class of physics problems relating to projectile motion. These objects have horizontal motion (labeled x) as well as vertical motion (labeled y). The velocity of such a projectile is a vector, having at any instant a magnitude and a direction. It is traditional to define the initial direction by the angle θ between the initial velocity vector and the horizontal axis. Then the components of the initial velocity vector, v_{xo} and v_{yo}, are given by the formulas shown below in the first row. In the idealized case with no air resistance, as the ball flies through the air, the x component of the velocity (v_x) remains constant while the y component of the velocity (v_y) changes in exactly the same way as it does in free fall. This is because the acceleration acts only in the vertical direction thereby affecting only the y component of the velocity and not the x component of the velocity.

Formulas that you should be familiar with:

$$v_{xo} = |v_o|\cos\theta \qquad v_{yo} = |v_o|\sin\theta$$

$$v_x = v_{xo} \qquad v_y = v_{yo} - gt$$

$$x = x_o + v_{xo}t \qquad y = y_o + v_{yo}t - \tfrac{1}{2}gt^2$$

Reference topics: projectiles, projectile motion, vectors

Simulation Details

The ball has an adjustable initial speed and can be launched into the air at an adjustable angle (the Initial Direction slider). This angle is defined as the angle between the initial velocity vector and the x axis (ground) and will be measured in degrees (straight up = 90°). As the ball flies through the air, the components of the velocity vector will be shown on your computer screen. To allow you to view the vector history, tracking (snapshots on the screen at equal time intervals) can be enabled in this simulation. You can turn tracking on and off by clicking the appropriate buttons. Clicking Erase will erase all of the tracks on the screen. This simulation will automatically pause when the ball is at the top of its flight. Just click on the Run button to go on from that point. If you want to go back to the beginning, click on Reset and then Run. This simulation will automatically stop when the ball is back near its initial y position and air resistance effects are ignored,

Set the initial speed to 40 m/s and the initial direction to 65°. Turn on tracking by clicking **On**. Click **Run** and observe the motion until the simulation pauses.

In what follows, circle the correct word.

- The magnitude of the y component of the velocity is (*remaining constant, increasing, decreasing*).
- The magnitude of the x component of the velocity is (*remaining constant, increasing, decreasing*).

Click **Run** to continue running and observe the motion on the way down.

In what follows, circle the correct word.

- The magnitude of the y component of the velocity is (*remaining constant, increasing, decreasing*).
- The magnitude of the x component of the velocity is (*remaining constant, increasing, decreasing*).

Click **Reset** and then click **Erase**. Set the initial speed to 36 m/s and the angle to 45°. This will make the x and y components of the initial velocity equal to each other.

Calculate the time of flight from the values of the initial speed and the angle.

time of flight = ____________

Click Run and observe the total motion of the ball.

From the simulation data, record the following:

time of the total flight = ___________

horizontal distance traveled = ___________

How does your calculation compare with the time that you recorded from the simulation?

Set the initial speed to 40 m/s and run for different angles. Record the data below. Time is the total time of flight and range is the final x position.

Time of Flight and Range vs. Angle

angle	10°	20°	30°	40°	45°	50°	60°	70°	80°
time									
range									

Plot time (vertical axis) vs. initial angle (horizontal axis) initial angle. You can do this by hand, or use a spreadsheet or graphing program if you have one on your computer.

time

initial angle

Plot range (vertical axis) vs. initial angle (horizontal axis).

range

initial angle

Are the graphs what you expected?

Are they straight lines or curves?

Are they symmetric about some point? Why or why not?

Self-Test Questions for Simulation 9

The following statements refer to a projectile launched at an angle other than 90°. True or false?

1. The velocity at the top of the flight is zero.
2. For a given initial speed, a larger angle results in a longer flight time.
3. For a given initial speed, a larger angle results in a longer range.
4. The x component of the velocity is constant throughout the entire motion.
5. The speed of the projectile decreases on the way up and increases on the way down.

SIMULATION 10 PROJECTILE MOTION WITH AIR RESISTANCE

Physics Review

The projectile and free fall problems that we have covered so far have been idealized. When real objects are thrown into the air, there are other forces causing changes in velocity besides the force of gravity. One of these other forces is the force due to air resistance. Solving projectile problems with air resistance is typically beyond the scope of an introductory physics class because the mathematics involved is more advanced. But with the aid of *Interactive Physics* you will be able to explore how air resistance affects the motion of a projectile. In this model air resistance will produce a force that is proportional to the square of the velocity of the ball and always opposite in direction to the velocity. The faster the ball is moving, the greater the acceleration (or deceleration if you prefer) and the more rapidly the ball will slow down. Unlike the acceleration due to gravity, which always acts downward whether the ball is moving up or down, the acceleration due to air resistance acts downward when the ball is moving upward and upward when the ball is moving downward (for one-dimensional motion). The force is also dependent on the size and shape of the ball and the material it is made from. A Ping-Pong ball will be more strongly affected by air resistance during flight than a steel ball of the same size. When an object is falling, at some point the acceleration due to gravity and the acceleration due to air resistance are very nearly equal and opposite. At this point the net acceleration is (for all practical purposes) zero and the ball will continue to fall (until it hits something) with a constant velocity. This velocity is known as the terminal velocity. Since wee are including air resistance the ball is not in free fall.

Reference topic: air resistance, terminal velocity

Simulation Details

In this problem you will not be able to adjust the angle or initial speed of the ball. The initial speed is fixed at 13.2 m/s with v_{xo} = 8 m/s and v_{yo} = 10.5 m/s. The air resistance is already on and you cannot adjust it. The simulation will automatically stop after about 5 seconds have elapsed.

Turn on tracking by clicking [On]. Click [Run] and observe the motion until the simulation stops. Use the tape player to locate the point where the ball is at the top of its flight.

From the simulation data, record or calculate the following:

elapsed time = ____________

x component of the velocity = ____________

y component of the velocity = ____________

x position = ________

y position = __________

Use the [tape player] to find the point where the ball has returned to initial y position (as close as you can get). From the simulation data, record or calculate the following:

elapsed time = ____________

x component of the velocity = ____________

y component of the velocity = ____________

x position = ________

y position = __________

When a projectile is launched without air resistance, the situation is quite symmetrical. The time up to the top is the same as the time down to the original starting height. For the case including air resistance, this is not the case. Record the time up and the time down for this simulation.

time up = ________________

time down = ______________

Why are the times different? Why is the time to go up less than the time to come down?

With no air resistance, the magnitude of the velocity is the same at the same height whether the ball is on the way up or down. Use the [o_o] to locate the two frames where the height of the ball is 2 m (as close as you can get). Enter the simulation data into the following table.

Magnitude of the Velocity (up vs. down)

	ON THE WAY UP	ON THE WAY DOWN
height of the ball		
x component of the velocity		
y component of the velocity		
magnitude of the velocity		

In this case with air resistance, is the magnitude of the velocity the same at the same height regardless of whether it is moving up or down?

Is there a point where the y component of the acceleration is close to g ($-9.8\ \mathrm{m/s^2}$)? If so, where is it?

What is the value of the x component of the velocity when the ball is at the top?

About how long does it take for the ball to attain its terminal velocity?

Does air resistance affect any of the following? How?

the time of flight:

the maximum height the ball attains:

the horizontal distance traveled:

Now focus your attention on the graphs in the simulation. Both acceleration graphs start out with negative values and decrease (in magnitude) to values of zero. That means that each component of the velocity is approaching a constant.

What constant value is the x component of the velocity approaching?

What constant value is the y component of the velocity approaching?

Sketch the graphs of velocity shown in the simulation in the space below. Use a solid line (or colored pencil if you have one). On the same graphs, sketch the components of the velocity curves if there had been no air resistance (use a dotted line or another color).

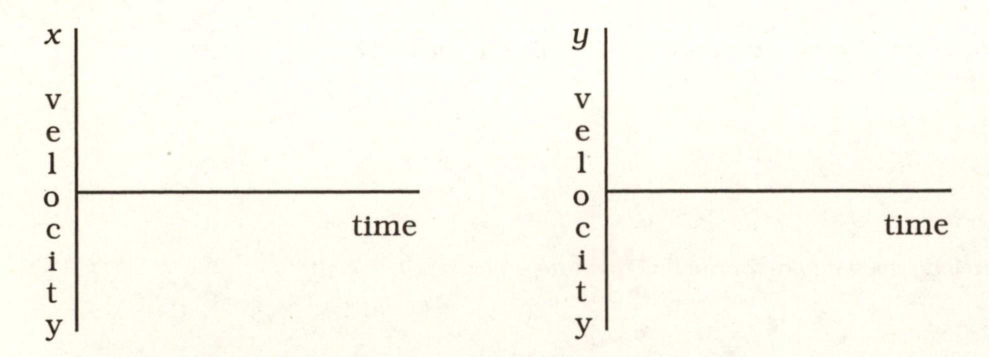

Sketch the graphs of position shown in the simulation in the space below. Use a solid line (or colored pencil if you have one). On the same graphs, sketch the position curves if there had been no air resistance (use a dotted line or another color).

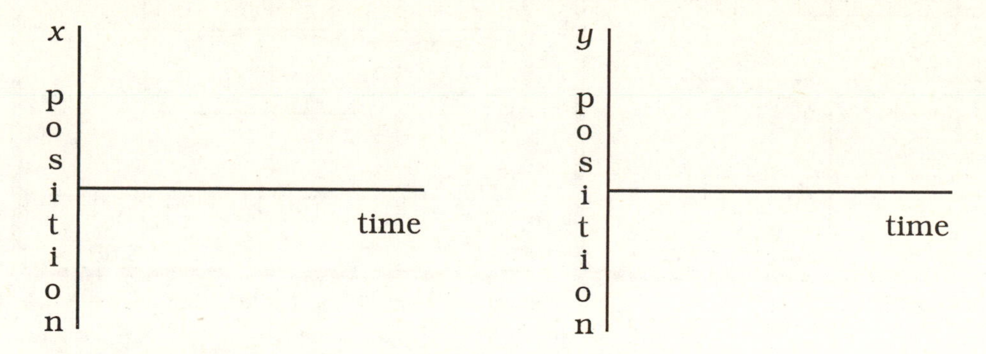

The graph of the x position is approaching a horizontal line. *What does that mean in terms of the physical path of the ball?*

The graph of the y position is also approaching a straight line (although not horizontal). *What does that mean in terms of the physical path of the ball?*

Self-Test Questions for Simulation 10

These questions refer to the graphs on the following page. Each of the six graphs has two numbered curves; one represents the motion of a ball thrown without air resistance, and the other represents the ball's motion with air resistance.

1. Which six curves (one from each graph) correspond to the motion with air resistance?
2. On the graph of the y component of velocity, which curve represents the ball that takes longer to reach its maximum height?
3. Which ball has fallen (y position) farther after 2.5 seconds have elapsed?
4. Which ball will have fallen (y position) farther at 3.5 seconds?
5. Which ball has gone farther (x position) in 2 seconds?

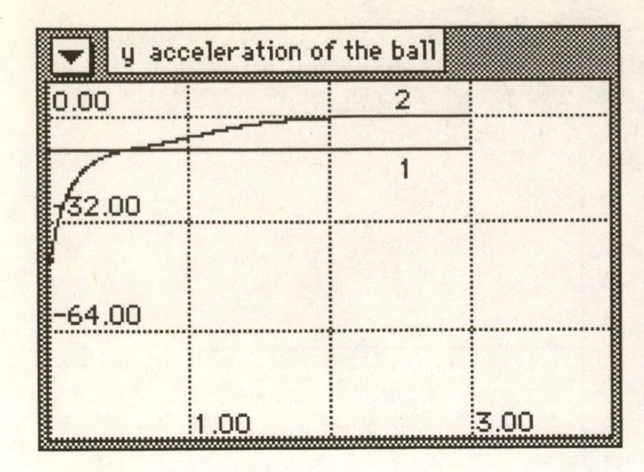
y acceleration of the ball
0.00
-32.00
-64.00
1.00
3.00
1
2

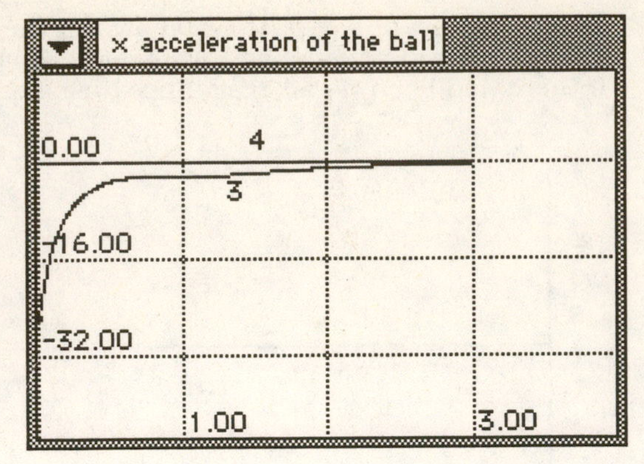
x acceleration of the ball
0.00
-16.00
-32.00
1.00
3.00
3
4

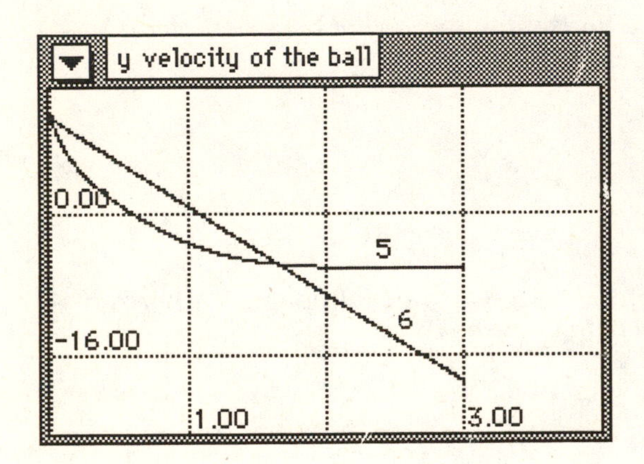
y velocity of the ball
0.00
-16.00
1.00
3.00
5
6

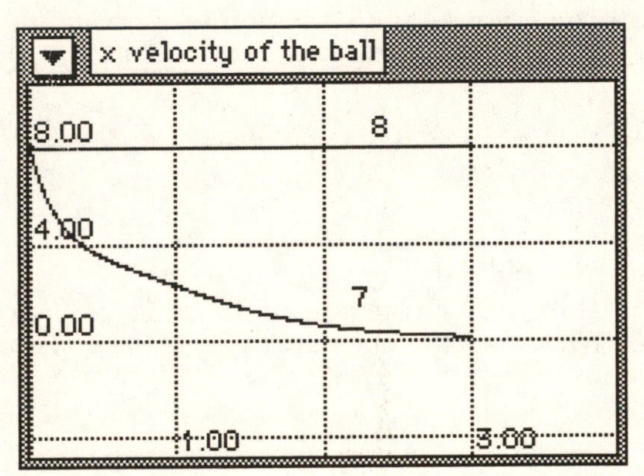
x velocity of the ball
8.00
4.00
0.00
1.00
3.00
7
8

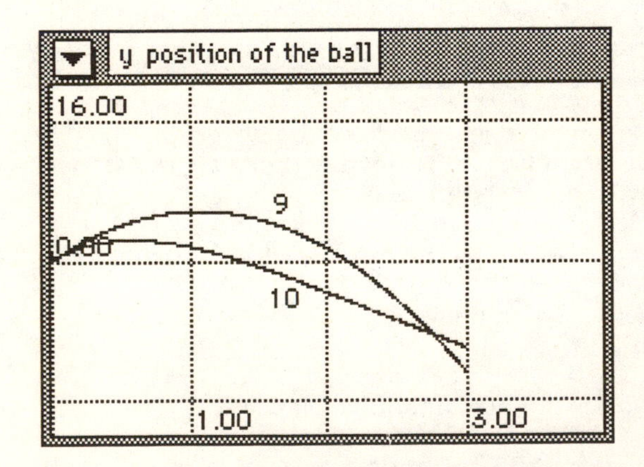
y position of the ball
16.00
0.00
1.00
3.00
9
10

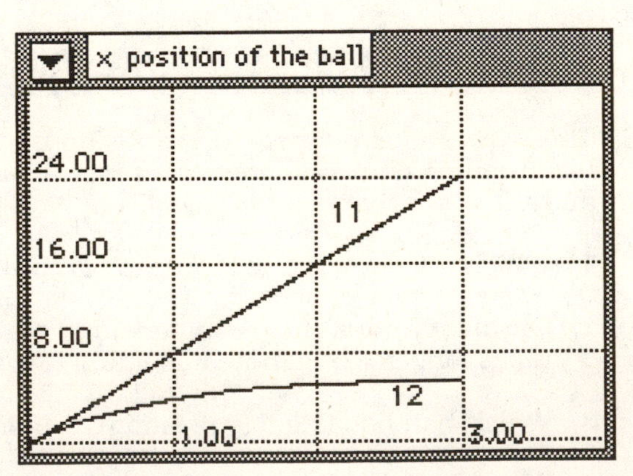
x position of the ball
24.00
16.00
8.00
1.00
3.00
11
12

SIMULATION 11 PROJECTILES AND RELATIVE VELOCITY

Physics Review

If you have ever been on a train, looked out the window, and seen another train moving backward, you already know that the motion of an object can depend on the observer and her motion relative to the object. You, sitting on the train, look at the person across the aisle and conclude that the person is not moving. You, sitting on the train moving at 65 mi/hr (about 30 m/s), look at the train outside your window (moving at 50 mi/hr) and conclude that it is moving backward. In fact, if you had a watch and knew how long the train cars were, you would observe that the train was moving backward at a speed of 15 mi/hr (about 6.7 m/s). Of course, the man sitting by the railroad tracks watching them both go by would say that neither is moving backward, that they both are moving forward, although one is moving faster than the other (65 mi/hr and 50 mi/hr respectively). You, sitting on the train, are observing the motion from the reference frame of the train. The man sitting by the tracks is observing motion from the reference frame of the Earth. Which one of you is correct in your observations? The answer, as we shall see, is that you both are. What the motion of an object looks like depends on the observer and her velocity relative to the object. In this problem we will study a ball thrown straight up from inside a moving car (a convertible with the roof down) and how that motion looks to observers in other cars.

Formulas you should be familiar with ($g = +9.8\ \text{m/s}^2$):

$$x = x_o + v_{xo}t \qquad y = y_o + v_{yo}t - \tfrac{1}{2}gt^2$$

$$v_x = v_{xo} \qquad v_y = v_{yo} - gt$$

Reference topics: relative velocity, reference frames, projectile motion

Simulation Details

In this simulation there are three cars moving along a road. A person in a sports car throws a ball straight up with a speed of 10 m/s. There is an old car that can move with a constant speed to the right or to the left, but the sports car and a hot rod can only move to the right. Each of the four buttons when clicked will change the reference frame to that of the object indicated by the name on the button. The "eye" will indicate which frame you are in. You can run the simulation all the way through, then reset and change the reference frame and run again, or you can change the reference frame in the middle of running. When you are in a particular frame of reference, say that of the hot rod, the path that you see on the screen is the path that the ball appears to have as seen by a person sitting in the hot rod. You will need to approximate position data by using the rulers on the left and bottom of the screen (put the mouse on the ball and read the values off the rulers.) This simulation will stop when the ball is back near its original height.

Set the velocity of the old car to 3 m/s, the velocity of the sports car to 5 m/s, and the velocity of the hot rod to 9 m/s. Click the button [Sports Car] to set the reference frame to that of an observer located in the sports car. Click [Run].

Where does the ball land?

Sketch the path of the ball.

From the simulation data, record or calculate the following:

time for the ball to complete its trip = ____________

how high the ball went (maximum y position – initial y position) = ____________

range of the projectile (according to the driver of the convertible) = ______________

Click [Reset]. Click the button [Ground] to set the reference frame to that of a stationary observer. Click [Run].

Where does the ball land?

Sketch the path that it took.

From the simulation data, record or calculate the following:

time for the ball to complete its trip = ____________

how high the ball went (maximum y position – initial y position) = ____________

range of the projectile (according to the stationary observer) = ________________

Would changing the reference frame to the hot rod affect any of the following?

time for the ball to complete its trip:

the height the ball will rise to:

the range of the projectile:

Predict what the path of the ball will be and sketch it in the area below.

Will it land in the sports car or not? Explain.

Click **Reset**. Click the button **Hot Rod** to set the reference frame to that of someone in the hot rod. Click **Run** and observe the motion until the simulation stops.

From the simulation data, record or calculate the following:

time for the ball to complete its trip = ____________

how high the ball went (maximum y position – initial y position) = ____________

range of the projectile (according to the driver of the hot rod) = ____________

Do you understand why the time of the flight is the same regardless of whether it is determined by an observer in the sports car, in the hot rod, or by the side of the road?

Do you understand why the height that the ball attains is the same for all four reference frames?

Do you understand why the range of the projectile is different for the different observers?

Click [Reset]. Click the button [Old Car] to set the reference frame to that of someone in the old car. Set the velocity of the old car to 3 m/s. Do not run yet.

Predict what the values of the following will be:

time for the ball to complete its trip:

the height the ball will rise to:

the range of the projectile:

Predict what the path of the ball will be and sketch it in the area below.

Click **Run** and observe the motion until the simulation stops.

From the simulation data, record or calculate the following:

time for the ball to complete its trip = ____________

how high the ball went (maximum y position – initial y position) = ____________

range of the projectile (according to the driver of the old car) = _______

Compare these values that you just recorded with your predictions on the previous page. If your calculation of range does not agree (approximately) with the value you get from looking at the simulation, you should read the following explanation very carefully.

Explanation

When calculating the range (horizontal distance, d), you use the formula $d = vt$. However, there are many velocities in this problem and you must use the correct one. In the previous case, the velocity that must be used is the velocity of the ball with respect to the old car. Since everything is in one dimension, the vector equations can be written as scalar equations and

$$v_{ball-oldcar} = v_{ball-sportscar} + v_{sportscar-oldcar}$$

which in this case gives $v_{ball-oldcar} = 0 + 2$ because

$$v_{sportscar-oldcar} = v_{sportscar-ground} - v_{oldcar-ground}$$

which in this case gives $2 = 5 - 3$

then

$d = vt$ gives a calculated range of 4 meters.

Assume that the old car is going the other way at the same speed (3 m/s). Calculate the range of the projectile now.

range = ___________

Sketch the path of the projectile in the space below.

Set the velocity of the old car to –3 m/s. Click **Run** and observe the motion to check your last calculation of range.

From the simulation data, record or calculate the following:

the range of the projectile = ___________

Compare this value to the value that you calculated at the top of this page. *Are they the same or different? Explain.*

Self-Test Questions for Simulation 11

These statements apply to the set of cars described in the simulation. True or false?

1. The time of flight of the ball is the same for all observers.
2. The ball always lands in the sports car.
3. The velocity of the ball at the top of the flight is the same for all observers.
4. Only the people in the sports car can say that $d = vt$.
5. The path the ball takes depends on the observer and her velocity with respect to the sports car.

SIMULATION 12 THREE FORCES ON A PUCK

Physics Review

Newton's first law tells us that a body will remain at rest or moving at a constant velocity so long as the net force on it is zero. Such a body is said to be in equilibrium. To determine whether or not a body is in equilibrium, one must add up all the forces acting on that body and see if the total sum is zero. Forces are vector quantities, specified fully by a magnitude and a direction. A body that is acted upon by only one (nonzero) force cannot be in equilibrium. There are many ways that the net force on an object can be zero. For example, two forces with the same magnitude and opposite direction will give rise to a state of equilibrium, and when three forces are involved (in two dimensions), things get even more interesting. In this problem we will investigate various ways for equilibrium to be achieved for a hockey puck to which three forces are applied.

A formula you should be familiar with:

$$F = ma$$

Reference topics: Newton's laws, forces, vectors

Simulation Details

In this simulation a hockey puck is viewed from above. Three forces act on the puck through its center. One of the forces, force 3, has a fixed magnitude and direction. It is shown in red and you cannot adjust it at all. The other two forces can be adjusted by you. Just type the number of each component in the corresponding text box. The largest value that you can type in these boxes is 40 and the smallest is –40.

Do not adjust any of the values in the simulation. Click Run and observe the motion.

Is the puck in equilibrium?

From the simulation data, record or calculate the following:

the horizontal acceleration of the puck = _________

the vertical acceleration of the puck = ___________

the magnitude of the acceleration of the puck = _________

the net force in the x direction = _______________

the net force in the y direction = _______________

the net force on the puck = _____________

the mass of the puck = ______________

Now you will try to get the puck to be in equilibrium. That requires that the net force on it be exactly zero. Try to do this by adjusting only force 1. Calculate the components of force 1 and its resulting magnitude and direction. As is the convention in *Interactive Physics*, the direction should be given with respect to the positive x direction.

x component of force 1 = _______________

y component of force 1 = _______________

magnitude of force 1 = _________________

direction of force 1 = ____________

Click **Reset**. Set the x and y components of force 1 to the values that you just calculated. Click **Run**.

Is the puck in equilibrium?

Click [Reset]. Change force 2 so that it has a magnitude of 10 N and points directly to the left. Click [Run].

You should now see that the net x force on the puck is zero. Calculate the y component of force 1 needed so that the puck will be in equilibrium.

y component of force 1 = __________

Click [Reset]. Set the y component of force 1 to the value that you just calculated. Click [Run].

Complete the following by drawing the vector for force 1. (tail on the tip of force 2)

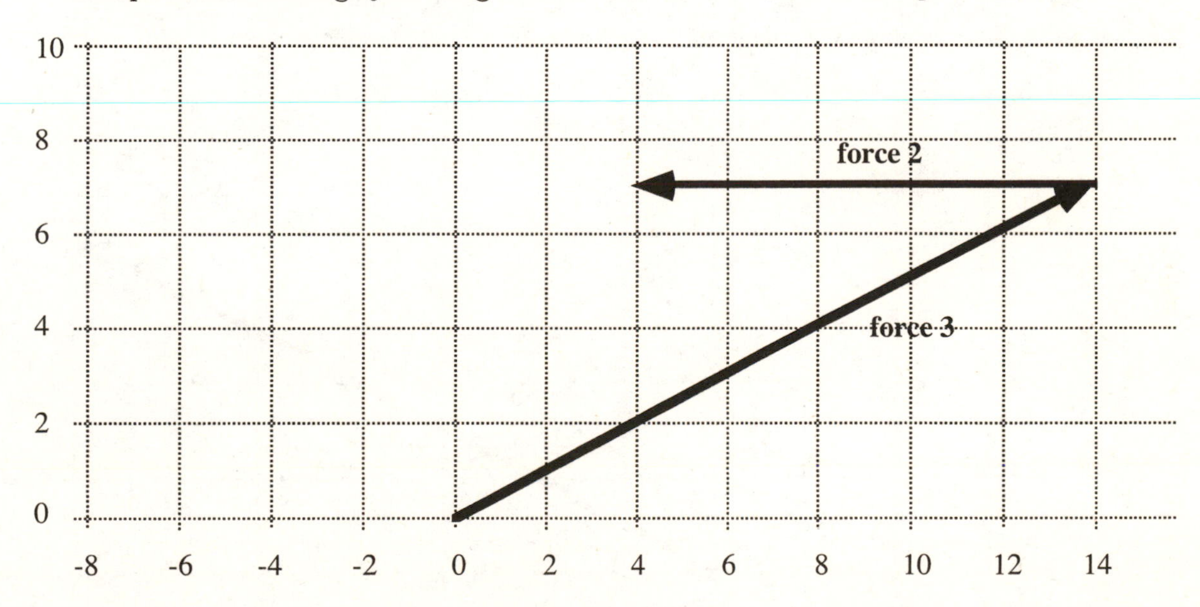

What does this show for the sum of the vectors (forces)?

Assume that the direction of force 1 is only the y direction (or negative y direction) and that the direction of force 2 is only the x direction (or negative x direction). Calculate the magnitudes of force 1 and force 2 so that the puck will be in equilibrium.

magnitude of force 1 = ____________

magnitude of force 2 = ____________

Click **Reset**. Set force 1 and force 2 to the values that you just determined.
Click **Run**.

Is the puck in equilibrium? Draw the forces on the puck as they are shown on your computer screen.

Complete the following by drawing the vectors for force 1 and force 2.

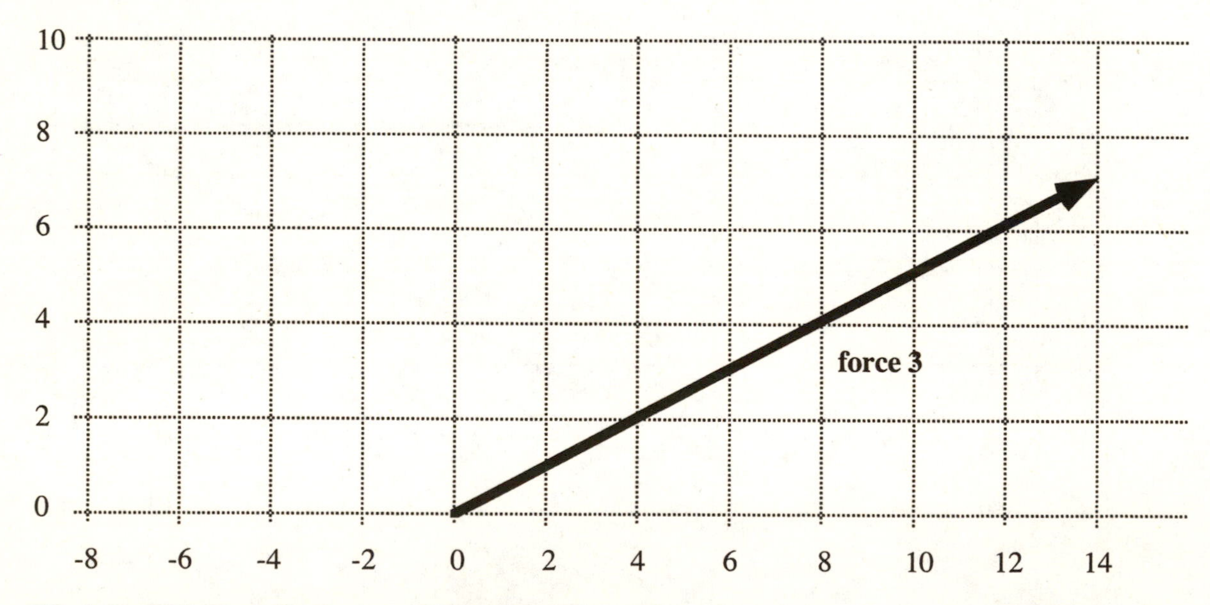

What does this show for the sum of the vectors (forces)?

Click **Reset**. Set the initial speed of the puck to 2 m/s. Set the x and y components of force 1 to 6 N and –7 N respectively and set the x and y components of force 2 to –20 N and 0 N respectively. Click **Run**.

Is the puck in equilibrium?

What is its acceleration?

What is the net force on it?

Is it moving?

Explanation

Even though the puck is moving, it is in equilibrium because the net force on it is zero. Objects can be stationary or moving at a constant velocity when they are in equilibrium.

In this part the goal is to get the puck to accelerate at 2.24 m/s² (a_x = 1 m/s², a_y = 2 m/s²). Given force 1, find force 2 needed for this acceleration (graphically). (tail on the tip of force 1)

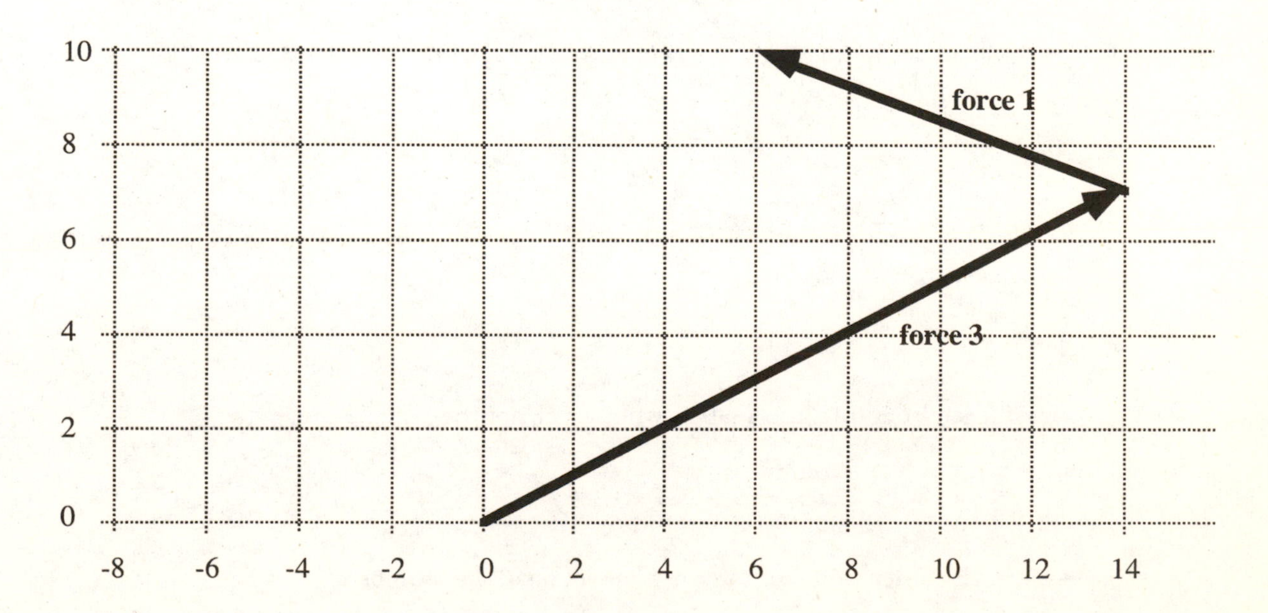

Click [Reset]. Set force 1 and force 2 to the values that will ensure this acceleration. Click [Run].

Did you get the proper acceleration? If not, try again. Hint: look at your graph on the previous page and read off the components of each force directly.

Self-Test Questions for Simulation 12

The questions apply to the three forces shown on the graph and applied to a puck as in the simulation.

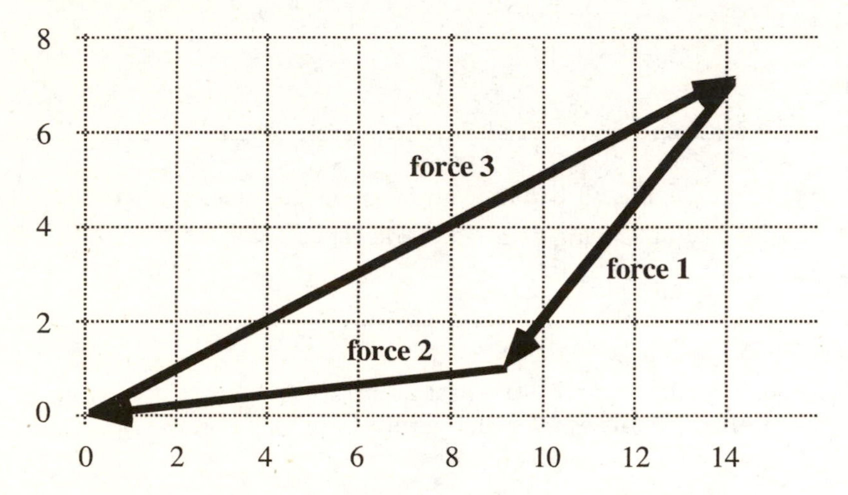

1. Label the following free body diagram for the puck with the forces 1, 2, and 3.

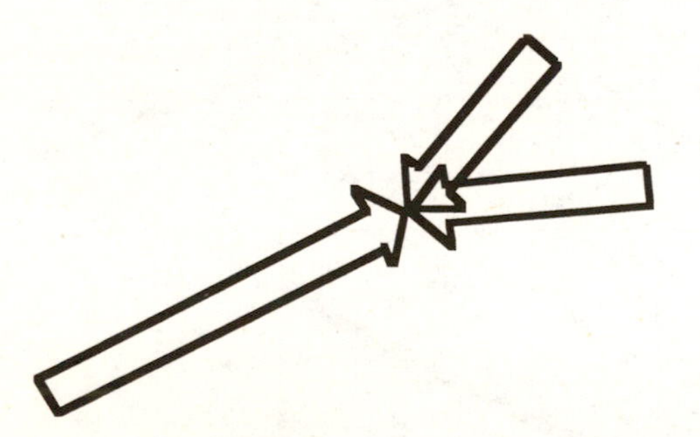

2. Can you tell from this information whether or not the puck is moving?

3. Is the puck accelerating?

4. Which force has the largest magnitude?

5. Is there any other value for force 1 that will result in equilibrium?

SIMULATION 13 HORIZONTAL MOTION WITH FRICTION (ENERGY)

Physics Review

When a puck slides on an ice rink, there is almost no friction present, which is good. When a large boulder rolls down a grassy hill, there is a lot of friction, which is also good. Depending on the physical situation and the goal, more or less friction may be needed. If a block is to slide on the floor with minimal loss of speed (or energy), then the coefficient of friction between the block and the floor must be small. In the real world we have minimal control over friction since it is determined by the complex interaction of the molecules on the surfaces involved. When one object is moving relative to another, the parameter of interest is called the coefficient of kinetic friction (μ_k), and the force of friction (*FF*) acting on an object of mass *m* is given by the normal force (*FN*) on that object times *m*. For the case of a block sliding on a horizontal surface, the normal force is equal to the weight of the block.

Formulas you should be familiar with:

$FF = \mu_k FN$ $\qquad W = FF \times distance$ $\qquad KE = \frac{1}{2}mv^2$

Reference topics: friction, normal force, work, kinetic energy

Simulation Details

The initial position of the block is the origin (0,0). You can adjust its mass and initial speed. The first section of the surface that it slides on is 4 meters long and has a coefficient of friction of zero. The second section, which is also 4 meters long, has an adjustable coefficient of friction. If you set the coefficient to a high value, corresponding to a very rough surface, the block will come to rest somewhere on that section. This simulation will stop when the block comes (close) to rest.

Set the coefficient of friction to 0.4 and set the mass of the block to 0.36 kg. Do not run yet.

Calculate the initial speed needed for the block to travel at least 2 meters into the second section.

initial speed = _______

Set the initial speed to 5 m/s and click Run. Observe the simulation until the block comes to rest.

Sketch the graphs of velocity, position and kinetic energy in the space below. Mark each graph with an X at the point where the block entered the region with friction.

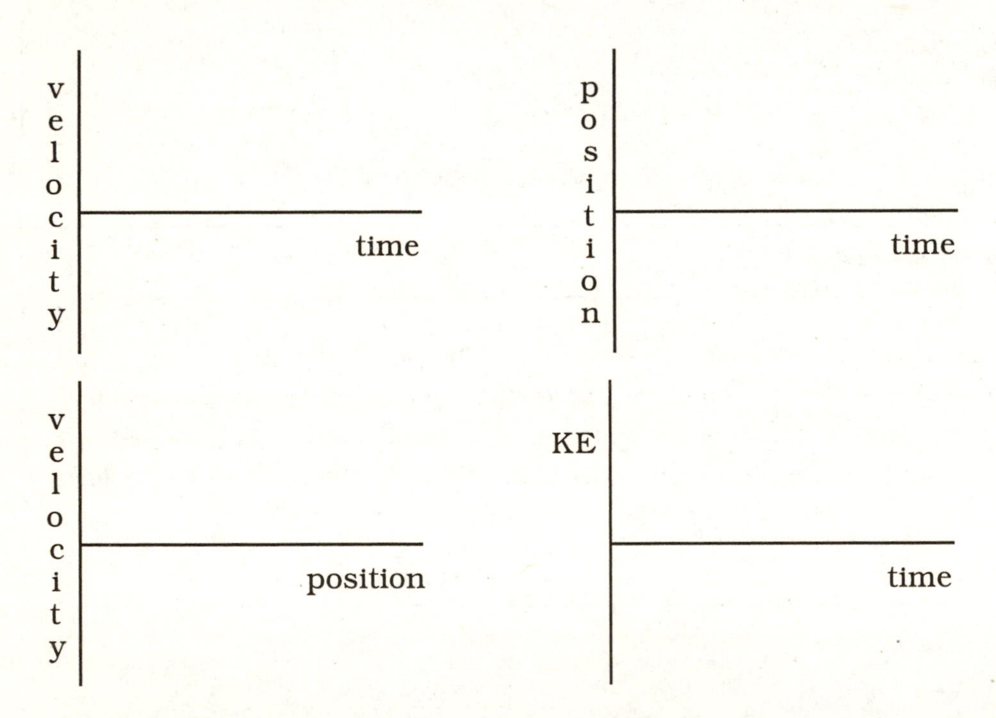

How do the two velocity graphs change when the block enters the area with friction? Why?

How does the position graph change when the block enters the area with friction? Why?

How does the kinetic energy graph change when the block enters the area with friction? Why?

From the simulation data, record or calculate the following:

the initial kinetic energy of the block = __________

the final kinetic energy of the block = __________

the energy lost by the block = __________

the total time to stop = __________

the frictional force on the block = __________

the position of the block when it stops = __________

the work done by the frictional force = __________

Predict which of the following will change if you change the mass to 0.18 kg. In what follows, circle the correct word. All statements are with reference to the previous case (m = 0.36 kg).

- The initial kinetic energy of the block will be (*larger, smaller, the same*).
- The final kinetic energy of the block will be (*larger, smaller, the same*).
- The kinetic energy lost by the block will be (*larger, smaller, the same*).
- The total time to stop will be (*larger, smaller, the same*).
- The frictional force on the block will be (*larger, smaller, the same*).
- The distance traveled by the block will be (*larger, smaller, the same*).
- The work done by the frictional force will be (*larger, smaller, the same*).

Click **Reset**. Set the mass to 0.18 kg and leave the initial speed set at 5 m/s.
Click **Run** and observe the motion until the block comes to rest.

From the simulation data, record or calculate the following:

CHECK COLUMN

______	the initial kinetic energy of the block = ________
______	the final kinetic energy of the block = ________
______	the kinetic energy lost by the block = ________
______	the total time to stop = ________
______	the frictional force on the block = ________
______	the position of the block when it stops = ________
______	the work done by the frictional force = ________

Compare these values to those that you recorded on the previous page. Put a check above next to the ones that you predicted correctly.

Now predict which of the following will change when you increase the coefficient of friction. In what follows, circle the correct word. All statements are with reference to the previous case ($\mu_k = 0.4$).

- The initial kinetic energy of the block will be (*larger, smaller, the same*).
- The final kinetic energy of the block will be (*larger, smaller, the same*).
- The energy lost by the block will be (*larger, smaller, the same*).
- The total time to stop will be (*larger, smaller, the same*).
- The frictional force on the block will be (*larger, smaller, the same*).
- The position of the block when it stops will be (*larger, smaller, the same*).
- The work done by the frictional force will be (*larger, smaller, the same*).

Click **Reset**. Leave the mass set to 0.18 kg and the initial speed set to 5 m/s. Change the coefficient of friction to 0.5. Click **Run**.

From the simulation data, record or calculate the following:
CHECK COLUMN

_______ the initial kinetic energy of the block = _________

_______ the final kinetic energy of the block = _________

_______ the kinetic energy lost by the block = _________

_______ the total time to stop = _______

_______ the frictional force on the block = _________

_______ the position of the block when it stops = _______

_______ the work done by the frictional force = _________

Compare these values to those that you recorded on the previous page. Put a check above next to the ones that you predicted correctly.

Optional:

- Vary the coefficient of friction and record data for the time spent on the surface with friction and the distance traveled on that section. Plot time vs. coefficient of friction and distance vs. coefficient of friction using a graphing program. Suggestion: use an initial velocity of 5 m/s and larger coefficients of friction to assure that the block will stop. Take at least 5 data points. *What do these graphs tell you?*

- Vary the mass of the block and record the total time to stop and the distance traveled on the second section. Plot time vs. mass and distance vs. mass using a graphing program. *What do these graphs tell you?*

- Vary the coefficient of friction for a constant mass and initial speed. *What do the two velocity graphs (v vs. t and v vs. x) tell you about the motion?*

Self-Test Questions for Simulation 13

The questions below refer to the graphs of the velocity and kinetic energy for two different blocks. Both blocks have the same initial speed of 5 m/s, and each slides the same distance on a frictionless surface before sliding along surfaces having different coefficients of friction.

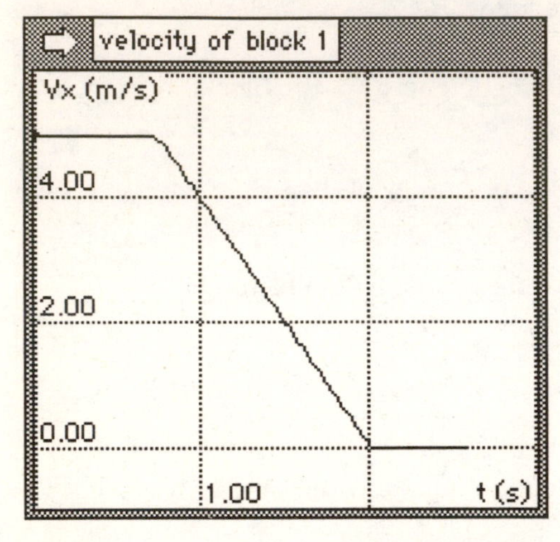

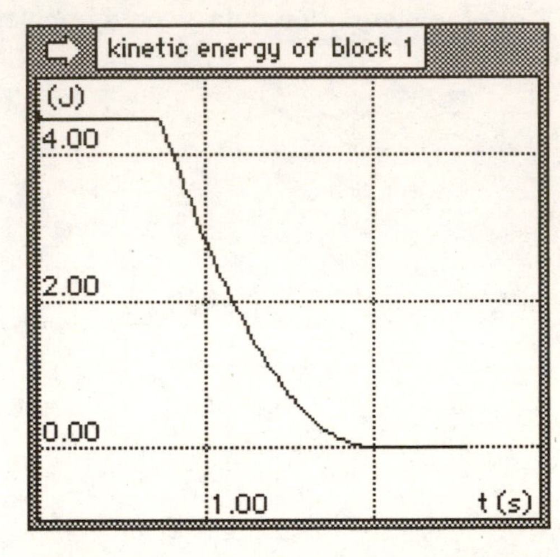

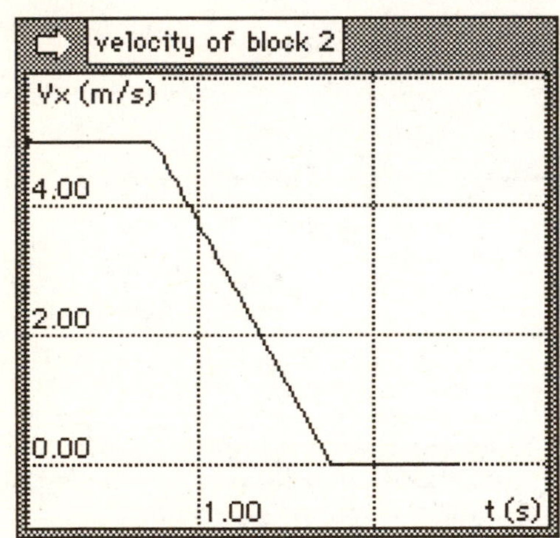

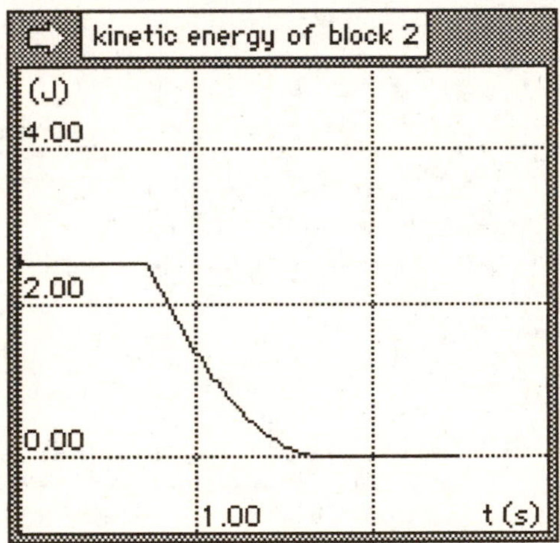

1. Which block has the smaller mass?

2. Which block loses more energy?

3. Which block requires more work to stop it?

4. Which block's acceleration has the largest magnitude?

5. Which one takes longer to stop?

SIMULATION 14 A CRATE PULLED ALONG A FLOOR

Physics Review

When a crate is pushed (or pulled) along a horizontal surface without friction by a constant force, it will always accelerate. The magnitude of the acceleration depends not only on the magnitude of the force, but also on its direction. For a fixed magnitude the maximum acceleration will occur when the force is applied parallel to the floor. If the force is applied to a crate of mass m at an angle as shown in the diagram labeled CASE 1, the acceleration will be given by $a = F\cos\theta/m$. As well, if the force is applied over and downward as is shown in the diagram labeled CASE 2, the acceleration will also be $a = F\cos\theta/m$ and the motion will be exactly the same. When the surface is not frictionless, the symmetry is broken and the motion in the two cases will be different. Although the component of the external force is the same in each case, the force of friction and therefore the net force will be different. The force of friction depends on the normal force, given by the formulas shown below.

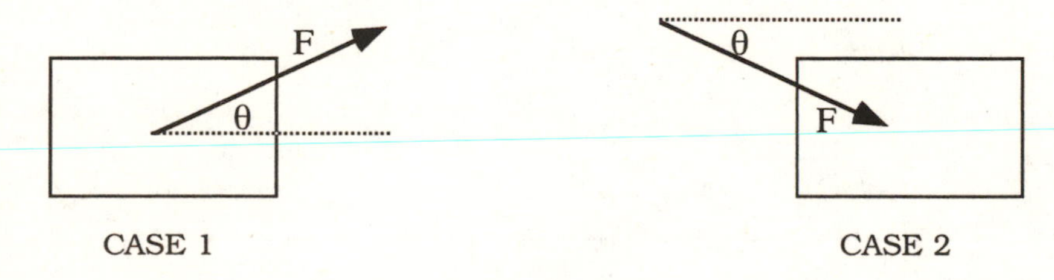

CASE 1 CASE 2

$FN = mg - F\sin\theta$ $FN = mg + F\sin\theta$

Formulas you should be familiar with:

$FF = \mu_k FN$ $W = FF \times distance$ $KE = \frac{1}{2}mv^2$

Reference topics: friction, normal force, work, kinetic energy

Simulation Details

The mass of the crate is fixed at 2 kg but its initial speed is adjustable. You can adjust the magnitude and direction of the external force that is applied to the crate, as well as the coefficient of friction between the crate and the floor. For this simulation positive angles (on the slider) will result in forces exerted at an angle as shown in CASE 1, and negative angles will result in forces exerted as shown in CASE 2. Output meters give simulation data for displacement, speed, acceleration and kinetic energy of the crate. The normal force between the crate and the floor is shown both as a vector and as a numerical output. This simulation will pause when the crate has been pushed a distance of about 4 meters and again when it has gone about 8 meters. **You will need to record data at both of these pauses.**

Assume a force of magnitude 5 N and no friction. Calculate the acceleration of the crate for these three cases.

$\theta = 0^\circ$

acceleration = ______

$\theta = 36^\circ$

acceleration = ______

$\theta = -36^\circ$

acceleration = ______

Set the magnitude of the force to 5 N and its direction to 36°. Set the coefficient of friction to 0.0. Click 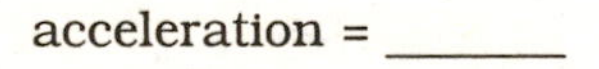

. Fill the data into the table on the next page and then reset and run again with 0° and –36°. Fill that data into the table as well. You should not have to run very long in each case.

5 N Force Applied at Different Angles

Angle	Acceleration	Normal Force
36°		
0°		
–36°		

Compare the acceleration values in this table to the values that you just calculated on the previous page. *Do they agree?*

Which of the three cases produced the largest acceleration? Why?

Was the acceleration the same for +36° as it was for –36°?

Which of the three cases resulted in the largest normal force on the crate? Why?

Was the normal force the same for +36° as it was for –36°?

Which case (if any) produced a normal force equal in magnitude to the weight of the crate?

Click **Reset**. Set the force to 10 N at –36° and the coefficient of friction to 0.25. Set the initial speed of the crate to 1.6 m/s. Click **Run** and at each pause record the data into the table on the next page.

A Block Being Pushed by an External Force at –36°

pause	position	KE	FN	FF	acceleration
1					
2					

From the data in the above table, calculate the work done by the external force during the time between the two pauses (include sign).

work done by external force = ____________

From the data in the above table, calculate the work done by the force of friction during the time between the two pauses (include sign).

work done by the force of friction = ________

From the data in the above table, calculate the change in kinetic energy between the two pauses (include sign).

change in kinetic energy = ________

Does the net work done equal the change in kinetic energy? (Hint: It should.)

Click [Reset]. Leave the magnitude of the force set to 10 N, the initial speed of the crate set to 1.6 m/s and the coefficient of friction set to 0.25. Change the direction of the force to 36°. Click [Run] and observe the motion. At each pause fill the data into the following table.

A Block Being Pushed by an External Force at 36°

pause	position	KE	FN	FF	acceleration
1					
2					

From the data in the above table, calculate the work done by the external force during the time between the two pauses (include sign).

work done by external force = ____________

From the data in the above table, calculate the work done by the force of friction during the time between the two pauses (include sign).

work done by the force of friction = ________

From the data in the table on the previous page, calculate the change in kinetic energy between the two pauses (include sign).

change in kinetic energy = ________

Click Reset. Leave the magnitude of the force set to 10 N, the initial speed of the crate set to 1.6 m/s and the coefficient of friction set to 0.25. Change the direction of the external force to 0°. Click Run and observe the motion. At each pause fill the data into the following table.

A Block Being Pushed by an External Force at 0°

pause	position	KE	FN	FF	acceleration
1					
2					

From the data in the above table, calculate the work done by the external force during the time between the two pauses (include sign).

work done by external force = ____________

From the data in the table on the previous page, calculate the work done by the force of friction during the time between the two pauses (include sign).

work done by the force of friction = ________

From the data in the table on the previous page, calculate the change in kinetic energy between the two pauses (include sign).

change in kinetic energy = ________

Of the three cases (36°, –36°, 0°) that you just studied:

Which had the largest acceleration? Why?

Which had the greatest frictional force on the crate? Why?

Which had the least amount of kinetic energy at the second pause? Why?

Which case had more work done by the external force? Why?

Calculate the force needed to keep the block moving at a constant speed for each of the three cases. Assume a coefficient of friction of 0.25.

force needed at 0° = _______________

Click **Reset**. Set the initial speed to 5.1 m/s. Set the angle to 0° and the force equal to the value that you just calculated (use two digits after the decimal point to get an accurate result). Click **Run**.

Is the crate moving at a constant speed?

Calculate the force needed to maintain a constant speed if it is applied at 36°.

force needed at 36° = _______________

Click **Reset**. Change the initial speed to 4.1 m/s. Set the angle to 36° and the force equal to the value that you just calculated (use two digits after the decimal point). Click **Run**.

Is the crate moving at a constant speed? Does it matter what the initial speed is?

Calculate the force needed to maintain a constant speed when applied at –36°.

force needed at –36° = ____________

Click **Reset**. Set the angle to –36° and the force equal to the value that you just calculated (use two digits after the decimal point). Click **Run**.

Which case required the greatest force to maintain a constant speed? Does this make sense?

Click **Reset**. Set the force to 10 N, the initial speed of the crate to 4.1 m/s and try to find a set of circumstances that causes the crate to slow down even though you are applying a constant force to it.

Record the coefficient of friction, the angle of the applied force, and the acceleration of the crate (during the motion) for one of these cases.

coefficient of friction = ____________

angle = ____________

acceleration (during the motion) = ____________

Click **Reset**. Set the coefficient of friction back to 0.25 and try to find the direction of the applied external force that gives the maximum acceleration.

Record the coefficient of friction, the angle, and the acceleration of the crate for this case.

coefficient of friction = ____________

angle = ____________

acceleration = __________

Explanation (Optional)

The angle that gives the maximum acceleration depends on the coefficient of friction. In fact, it is given by $\theta = \tan^{-1} \mu_k$. This is not too difficult to derive if you know calculus (this is an optional discussion so you need read no further unless you are curious about how to derive this result). Assume a force applied as in CASE 1. The net force on the crate can be written as a function of F and θ.

$$F_{net} = F\cos\theta - \mu(mg - F\sin\theta) = ma$$

If you take the derivative of this function with respect to θ and then set that to zero, you will get the above result.

$$\frac{\partial F_{net}}{\partial \theta} = -F\sin\theta + \mu F\cos\theta = 0 \qquad \text{or} \qquad \tan\theta = \mu$$

Did your investigation lead to the same conclusion?

Self-Test Questions for Simulation 14

The following statements all refer to a crate that has an initial speed of 2 m/s and is being pulled by an external force along a horizontal surface with friction. True or false?

1. The kinetic energy of the crate is always increasing as long as the external force is being applied.
2. The work done by the external force is always equal to the work done by the frictional force.
3. When the external force has a fixed direction and $0^\circ < \theta < 90^\circ$, increasing the magnitude of the force will always result in a larger acceleration.
4. An angle of 45° results in a smaller force of friction than an angle of 36°.
5. An angle of 30° results in a larger normal force than an angle of -30°.

SIMULATION 15 A BLOCK AND MASS ATTACHED WITH A ROPE OVER A PULLEY

Physics Review

When a problem involves multiple objects connected together by a constraint such as a rope, we take the rope to be massless and inextensible and analyze the forces on each of the objects separately. It is important to choose a positive direction and to define all vector quantities as positive or negative with respect to that direction for the entire problem. We are free to choose this positive direction and the reference point for zero gravitational potential energy as well. We will denote the tension in the rope by T. In this problem we have a block on a table connected by a rope over a pulley to another hanging block. We will choose the negative direction to be to the left and then over the pulley and down. Then, for the block on the table, a negative velocity will mean that it is moving to the left, and for the hanging block, a negative velocity will mean that it is falling.

Formulas you should be familiar with:

$FF = \mu_k FN$ $F = ma$ $PE = mgy$

Reference topics: Atwood Machine, pulleys, friction

Simulation Details

The two blocks each have an adjustable mass. The coefficient of friction between block 1 and the tabletop is adjustable, as is the initial speed of the system (e.g., either from rest or with a small initial speed). Outputs will allow you to determine the tension in the rope as well as the acceleration and energy of the two blocks. Zero potential energy is taken to be at the location of the center of the block on the table (y = 0.0 m).

Draw all the forces on the two blocks. Denote the mass of the block on the table by m_1 and the mass of the hanging block by m_2.

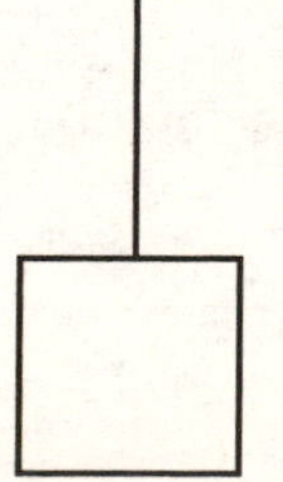

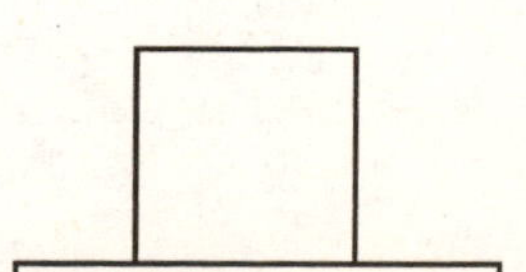

Use these free body diagrams to help derive the following expression for the acceleration of block 1. (Note that it is the same as the acceleration of block 2!)

$$a = \frac{(m_2 - \mu_k m_1)g}{m_1 + m_2}$$

Calculate what the acceleration will be if the masses are equal and the coefficient of friction is zero.

acceleration = ______________

Set the mass of both blocks to 2 kg and the coefficient of friction to zero.
Click Run.

From the simulation data, record the value of the acceleration of the blocks and the tension in the rope.

acceleration = ______________

tension = ______________

From the simulation data, record or calculate the following:

the total initial energy of block 1 = ______________

the total initial energy of block 2 = ______________

the total final energy of block 1 = ______________

the total final energy of block 2 = ______________

Is the mechanical energy of block 1 conserved?

Is the mechanical energy of block 2 conserved?

Is the mechanical energy of the system conserved?

What is the largest value that the acceleration can have in <u>this simulation</u>? What would be the values of m_1, m_2 and μ for that case?

Now you will explore cases <u>with friction</u>. Assume a coefficient of friction of 0.25 and a mass of 4 kg for block 1. Calculate the mass of block 2 needed to assure that the system will not accelerate [illegible] calculate the resulting tension in the rope.

m_2= __________

tension = __________

Click **Reset**. Set the mass of block 1 to 4 kg, the coefficient of friction to 0.25 and the mass of block 2 to the value that you just calculated. Set the initial speed of the system to 0.4 m/s. Click **Run** and observe the motion.

From the simulation data, record the value of the acceleration of the blocks and the tension in the rope.

acceleration = ____________

tension = ____________

Is the system not accelerating? Are the blocks moving?

Click **Reset**. Keep the mass of block 1 set to 4 kg and set the mass of block 2 to 3 kg. Set the initial speed of the system to 0.0 m/s. Vary the coefficient of friction and record the simulation data into the table.

Acceleration and Tension vs. Coefficient of Friction

coefficient of friction	acceleration of system	tension in the rope
0.00		
0.10		
0.20		
0.40		
0.65		
0.75		
0.85		
0.90		

Explanation

Note that the maximum value of the tension is equal to the weight of the hanging block and the tension has that value so long as the coefficient of friction is equal to or greater than 0.75 (the ratio of the mass of the block 2 to the mass of block 1).

From the data in your table plot the following:

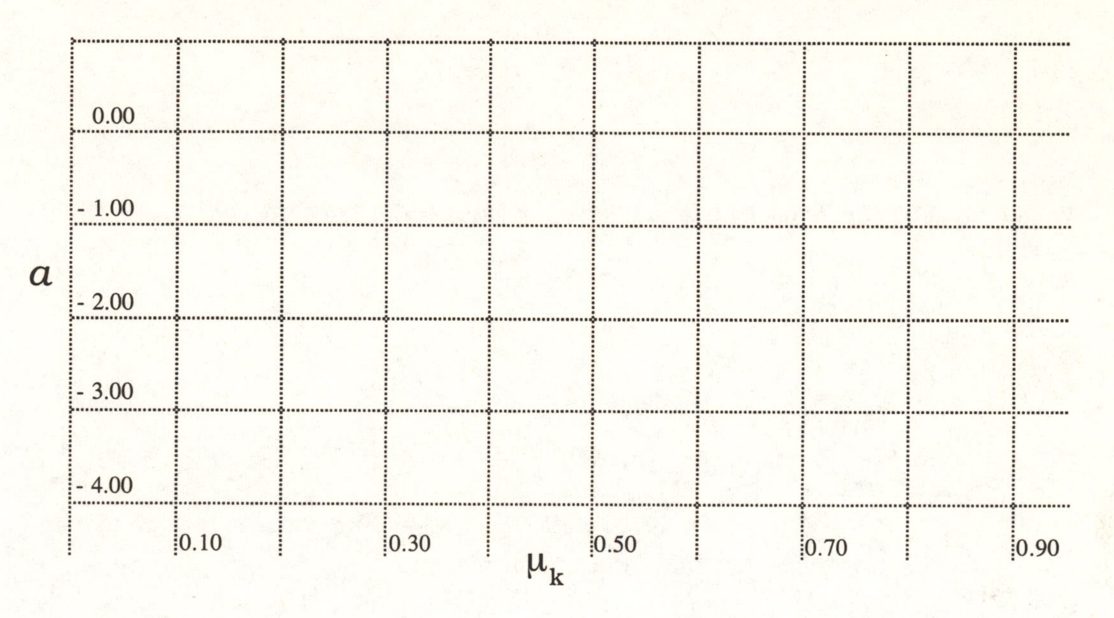

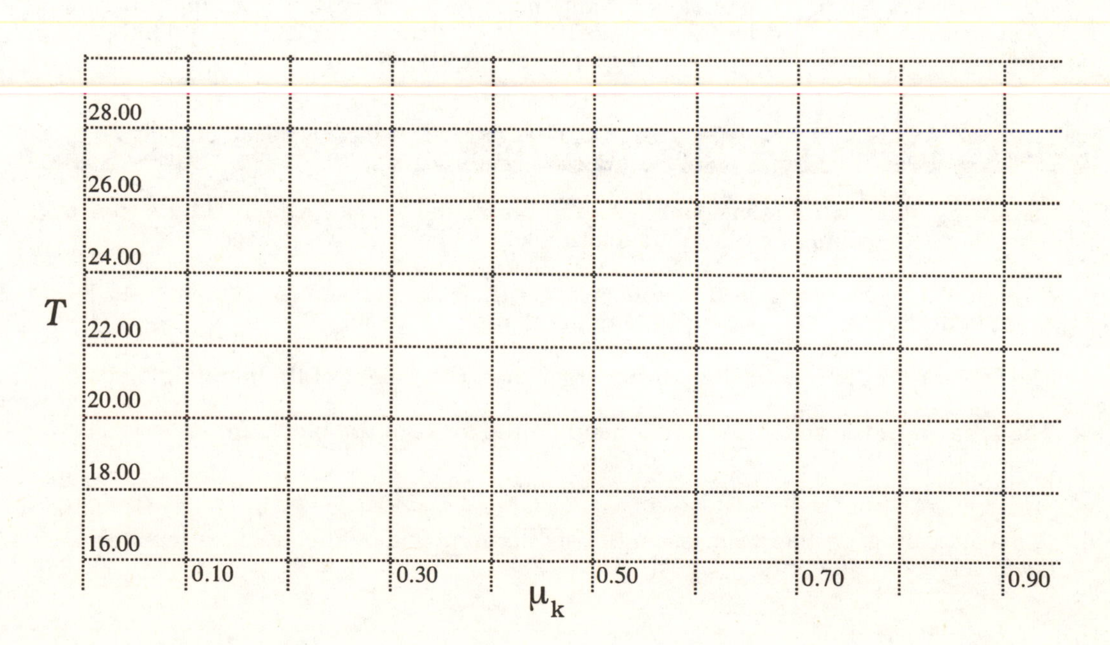

Think about how the graphs would be different if the mass of block 1 was the same (4 kg), but the mass of block 2 was only 2 kg.

Would the maximum tension be the same, more or less?
Explain.

Would the maximum tension occur at a higher or lower coefficient of friction?
Explain.

Click Reset. Set the mass of block 2 to 2 kg and the mass of block 1 to 4 kg. Try different values for the coefficient of friction.

Optional: Set the mass of block 1 to 4 kg and the mass of block 2 to 3 kg again. This time, however, set the initial speed of the system to 0.4 m/s and run for the same values of coefficient of friction that you did in the table on page 86. Record data on both the initial acceleration and tension and the eventual acceleration and tension for each case. *Why are things different if the block is initially moving?* If you have a graphing program, you can plot the initial acceleration vs. coefficient of friction and the initial tension vs. coefficient of friction and see that both graphs are straight lines.

Self-Test Questions for Simulation 15

The following statements refer to a block on a horizontal surface attached by a rope over a pulley to another hanging block. True or false?

1. If the mass of the block on the table is larger than the mass of the hanging block and there is no friction, the system will not accelerate.

2. The tension in the rope will never be larger than the weight of the hanging block.

3. The tension in the rope does not depend on the mass of the block on the surface.

4. It is not possible for the hanging block to move upward.

5. For fixed values of the two masses, the acceleration will be the smallest when the coefficient of friction is the largest.

SIMULATION 16 A BLOCK MOVING UP A HILL WITHOUT FRICTION

Physics Review

When an object, such as a block, is placed on an inclined (frictionless) plane, it will slide down with a constant acceleration. The magnitude of this acceleration depends on the angle of inclination (θ), which is measured with respect to the x axis. The magnitude of the acceleration is given by the formula shown below and its direction is always down the hill. Even though this appears to be a two-dimensional problem (since we could speak of x and y components), it can easily be treated as motion in one dimension with the direction defined as up or down the plane. Then the same equations for position, velocity, and acceleration for one-dimensional motion with constant acceleration will be valid.

$a = -g\sin\theta$ $\quad v = v_o - (g\sin\theta)t$ $\quad d = v_o t - \frac{1}{2}(g\sin\theta)t^2$

By writing these equations in this way, we are taking the positive direction to be up (parallel to) the hill and $g = +9.8\ \text{m/s}^2$.

Reference topics: inclined planes, constant acceleration

Simulation Details

The angle of inclination of the hill is fixed at 30°. When you run the simulation, a block will be projected up the hill with an initial speed that you can adjust. You will also be able to adjust the mass of the block. The simulation will automatically stop when the ball has returned to (near) its original position. Vectors representing the velocity and acceleration of the block as well as the force of gravity and the normal force on the block will be displayed on your screen. Output meters of position, velocity, and acceleration will allow you to study their relationship.

Mathematics Help

$\sin 30^\circ = 0.500$ $\quad \cos 30^\circ = 0.866$

Assume the initial speed of the block is 9 m/s and its mass is 3 kg. Calculate the distance that the block will travel up the hill and the time it will take for the block to go that distance.

time = _______

distance = __________

What will be the acceleration of the block when it reaches its maximum distance up the hill?

What will be its velocity at that point?

Set the initial speed to 9 m/s, the mass to 3 kg and click **Run**. Observe the motion of the block and the vectors associated with it.

From the simulation data, record or calculate the following:

time to the top = ___________

distance traveled up the hill = ____________

acceleration at the top = __________

velocity at the top = ___________

How do these values compare with your calculations and answers on this page?

Describe what is happening to the following vectors when the block is moving up the hill. Be sure to indicate information about both magnitude and direction.

velocity:

acceleration:

force of gravity:

normal force:

net force:

Draw a free body diagram of the block (showing all the force and velocity vectors) when it is halfway up the plane.

Use the to go back and study the same vectors when the block is moving down the hill. Be sure to indicate information about both magnitude and direction.

velocity:

acceleration:

force of gravity:

normal force:

net force:

Draw a free body diagram of the block (showing all the force and velocity vectors) when it is halfway down the plane.

From the simulation data, record or calculate the following (for the initial v = 9 m/s):

distance the block traveled up the hill = __________

the total time it took for the trip = ________

the velocity of the block when it returns to the bottom = ________

acceleration = ____________

force of gravity = ___________

normal force = _________

Which of the above quantities are affected (and how) by changing the initial speed of the block? Check your answers using the simulation.

Which of the above quantities are affected (and how) by changing the mass of the block? Check your answers using the simulation.

Self-Test Questions for Simulation 16

The following statements refer to a block being propelled up a frictionless hill that is inclined upward at 20°. True or false?

1. The distance a block will travel up the hill depends on the initial speed of the block.
2. If the mass of the block is larger, the distance traveled up the plane will be less.
3. The smaller the mass, the less time it takes to travel 1 meter up the hill.
4. The net force on the block is never zero.
5. The speed of the block when it returns to the bottom is the same as its initial speed.

SIMULATION 17 A BLOCK MOVING ON A HILL WITH FRICTION

Physics Review

In the real world objects do not slide on frictionless surfaces. The force of friction (FF) between two surfaces moving relative to one another is given by the product of the coefficient of friction (μ_k) and the normal force (FN) and is always directed opposite to the direction of relative motion. For an object of mass m moving on a plane inclined at an angle θ, the normal force (FN) is given in turn by the formula shown below. Hence, in the direction of motion, at any time, the net force on the mass is given by a combination of $mg\sin\theta$ and $\mu mg\cos\theta$. This net force may be positive, negative, or zero. We will define the positive direction to be up the plane. In the previous simulation you found that there was symmetry in the case of an object moving up (and then down) on an inclined plane without friction. If you recall from a previous simulation on projectiles (Simulation 10), when air resistance was included, symmetry was broken. When friction is included in motion on an inclined plane, you will find that symmetry is broken as well.

Formulas you should be familiar with:

$FR = \mu_k FN$ $\quad\quad FN = mg\cos\theta$

$F_{parallel\ to\ plane} = -\mu_k mg\cos\theta - mg\sin\theta$, when the block is moving up the hill.

$F_{parallel\ to\ plane} = \mu_k mg\cos\theta - mg\sin\theta$, when the block is moving down the hill.

Reference topics: inclined planes, friction

Simulation Details

You can adjust the initial speed and mass of the block as well as the coefficient of friction between the block and the plane. The angle of inclination of the hill is fixed at 30°. When you run this simulation, the force of friction, gravity, and the normal force vectors will be shown as well as the velocity and acceleration of the block. Note that, on the velocity graph, positive values represent velocity vectors pointing uphill and negative values represent velocity vectors pointing downhill.

Set the initial speed of the block to 8 m/s, the mass of the block to 2 kg, and the coefficient of friction to 0.0. Click **Run** and observe the motion of the block up and back down the hill.

Use the simulation data to fill in the table below to get numbers for comparison.

Inclined Plane with No Friction

mass	accel - up	accel - down	time - up	time - down	distance - up

Sketch the velocity graph in the region below.

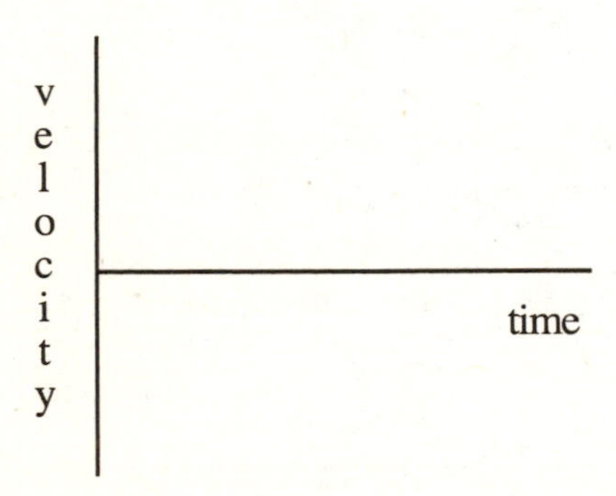

Click **Reset**. Keep the initial speed of the block at 8 m/s and the mass at 2 kg. Change the coefficient of friction to 0.2. Click **Run** and observe the motion of the block up and back down the hill.

Draw the free-body diagram of the block when it is about halfway up the hill and again when it is halfway down the hill. Label the forces of gravity (*FG*), friction (*FF*) and the normal force on the block (*FN*), as well as the velocity (v) and the acceleration (a) in each case.

Use the simulation data to fill in the table below.

Inclined Plane with Friction = 0.2

mass	accel - up	accel - down	time - up	time - down	distance - up

Sketch the velocity graph in the region below. Mark the point on the graph where the block is changing direction with an X.

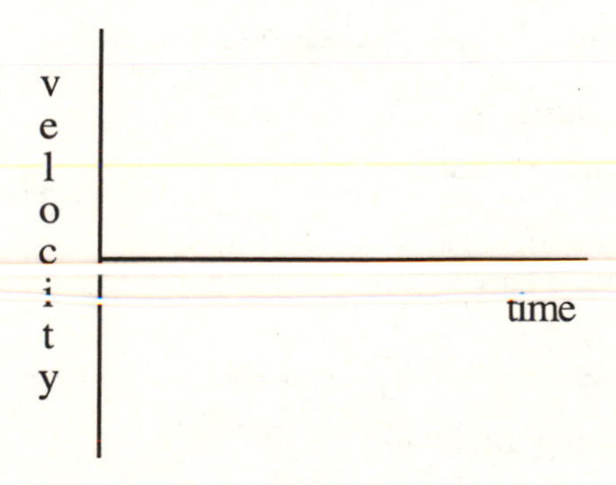

How is the velocity graph different? Why does the slope change?

Which quantities in the table above have changed from the case without friction, and how? (See table on page 94)

Now suppose that you were to increase the coefficient of friction to 0.4. In what follows, predict how things will change from the last case (friction of 0.2). Circle the correct words.

- The magnitude of the acceleration going up will (*be larger, be smaller, stay the same*).
- The magnitude of the acceleration going down will (*be larger, be smaller, stay the same*).
- The distance traveled by the block will (*be larger, be smaller, stay the same*).

Predict what the velocity graph will look like now and sketch it on the previous page. *How will it differ from the last case?*

Click **Reset**. Keep the initial speed of the block set to 8 m/s and the mass at 2 kg. Increase the coefficient of friction to 0.4. Click **Run**. Observe the motion of the block up and back down the hill.

Use the simulation data to fill in the table below.

Inclined Plane with Friction = 0.4

mass	accel - up	accel - down	time - up	time - down	distance - up

How do these values compare to your predictions at the top of this page?

Explanation

The time up is no longer equal to the time down as it was for the case with no friction. This was also true in the problem on projectiles with air resistance. The similarity is no coincidence. Both cases involve forces, acting opposite to the velocity of the object, that dissipate mechanical energy.

For the case with no friction, changing the mass of the block has no effect on any of the five quantities you recorded in your table. For the case with friction, predict how the following will or will not change if you decrease the mass from 2 kg to 1 kg.

In what follows, circle the correct words.

- The magnitude of the acceleration going up will (*be larger, be smaller, stay the same*).
- The magnitude of the acceleration going down will (*be larger, be smaller, stay the same*).
- The distance traveled by the block will (*be larger, be smaller, stay the same*).

Click **Reset**. Keep the initial speed at 8 m/s and the coefficient of friction at 0.4. Change the mass to 1 kg. Click **Run** and observe the motion.

Use the simulation data to fill in the table below.

Inclined Plane with Friction = 0.4

mass	accel - up	accel - down	time - up	time - down	distance - up

How do these values compare to your predictions at the top of this page?

Explanation

Even with friction, the motion is unaffected by changing the mass of the block. Also, in any case with friction, the average magnitude of the acceleration of the block is always equal to the acceleration of the block with no friction. As you increase the coefficient of friction, the acceleration of the block sliding down the hill is approaching zero.

Self-Test Questions for Simulation 17

The questions refer to these two sets of graphs of speed and position for two blocks that have been projected up a hill. The first set of graphs pertains to one block and the second set pertains to another. The blocks have the same mass.

BLOCK 1

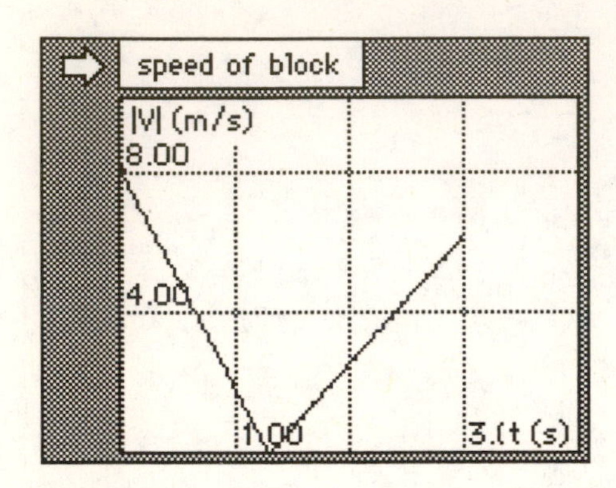

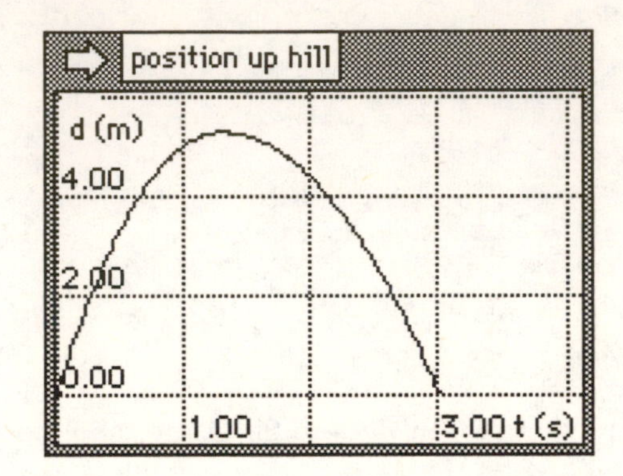

BLOCK 2

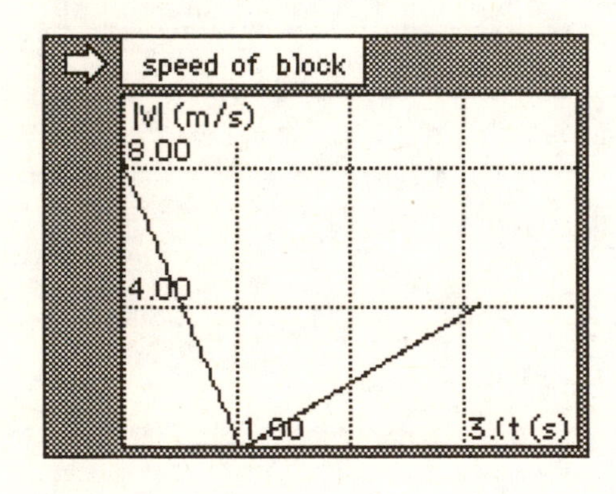

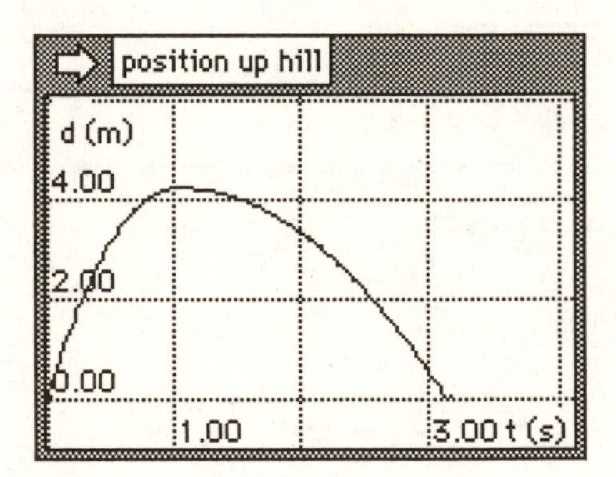

1. Which block had the greater initial speed?
2. Which block goes farther up the hill?
3. Which block takes longer to go up the hill?
4. Which block takes longer to come down the hill?
5. Which block experienced a greater force of friction?

SIMULATION 18 A BLOCK MOVING ON A HILL WITH FRICTION (ENERGY)

Physics Review

Inclined plane problems can also be solved and studied effectively by looking at the energy of the system. For motion on an inclined plane the mechanical energy of the block sliding on the plane is given at any instant by the sum of its kinetic and gravitational potential energy. For motion without friction this sum is always equal to a constant, and given the initial energy of the block, one can easily calculate how high up the plane it will go. One just has to get the geometry straight. We use

$$\tfrac{1}{2}mv^2 = mgh$$

to solve for h and then

$$d = \frac{h}{\sin\theta}$$

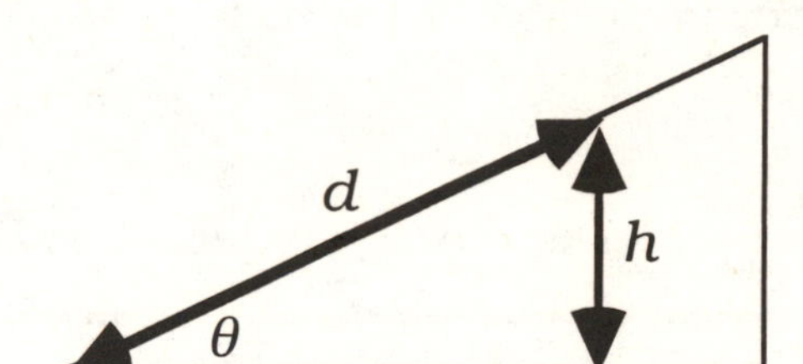

Once we involve friction, things are not as simple. Friction is a dissipative force, which means that mechanical energy is lost. However, we can still use energy methods to solve these problems. Mechanical energy will not be conserved, but we can use the fact that the work done by the frictional force (work = force x distance) must be subtracted from the kinetic energy and that amount is now equal to the potential energy. The same relationship still holds between h and d.

$$\tfrac{1}{2}mv^2 - \mu_k mg(\cos\theta)d = mgh$$

Reference topics: inclined planes, friction, energy

Simulation Details

This simulation is similar to the last one, only now you will study energy and the work done by the frictional force. The hill is inclined at 30°.

Assume that the initial speed of the block is 8 m/s. Use energy methods to calculate the distance the block will travel up a frictionless inclined plane.

distance up the plane = ___________

Set the initial speed of the block to 8 m/s, the mass of the block to 2 kg, and the coefficient of friction to 0.0. Click Run and observe the motion of the block up and back down the hill.

Use the simulation data to fill in the table below to get numbers for comparison.

Inclined Plane with No Friction

initial speed	distance - up	FF - up	work - up	FF - down	work - down

net-work	KE - initial	KE - final	ΔKE	% KE - lost	mass

Sketch the graphs of kinetic and potential energy in the region below.

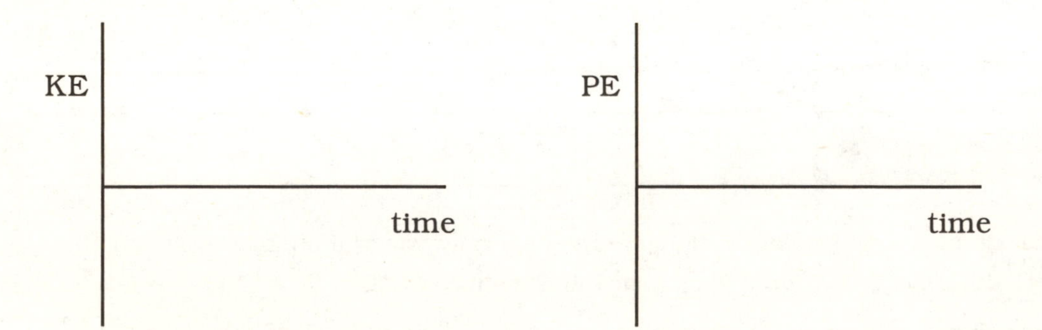

Assume that the initial speed of the block is 8 m/s. Use energy methods to calculate the distance the block will travel up the plane if the coefficient of friction is 0.2.

distance up the plane = ___________

Click **Reset**. Keep the initial speed of the block at 8 m/s and the mass at 2 kg. Change the coefficient of friction to 0.2. Click **Run** and observe the motion of the block up and back down the hill.

Use the simulation data to fill in the table below. Round the numbers in the second row to whole numbers.

Inclined Plane with Friction = 0.2

initial speed	distance - up	FF - up	work - up	FF - down	work - down

net-work	KE - initial	KE - final	ΔKE	% KE - lost	mass

Check to see that the work done by the force of friction is equal to the kinetic energy lost.

Sketch the graphs of kinetic and potential energy in the region below. Place an X at the point where the block is reversing direction. Indicate that time on the graphs.

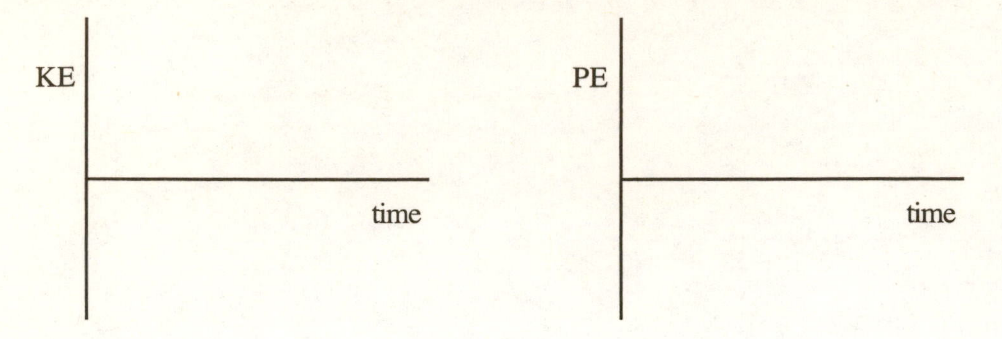

Why are the graphs no longer symmetric?

Predict what the graphs will look like if you double the coefficient of friction. Sketch them in the region below. Place an X at the point where the block is reversing direction.

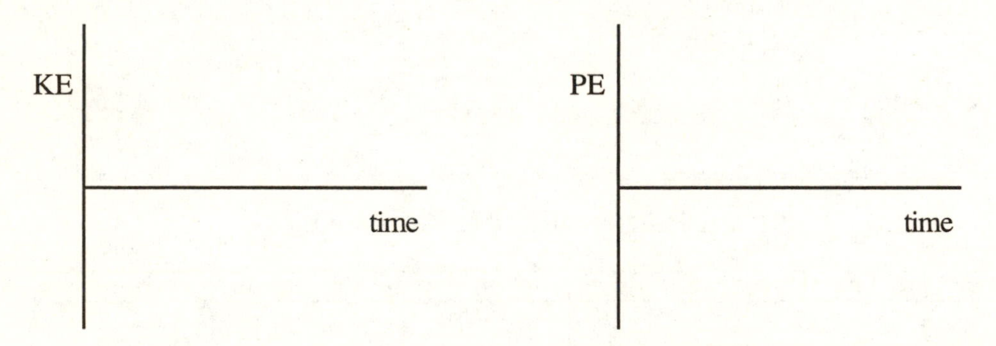

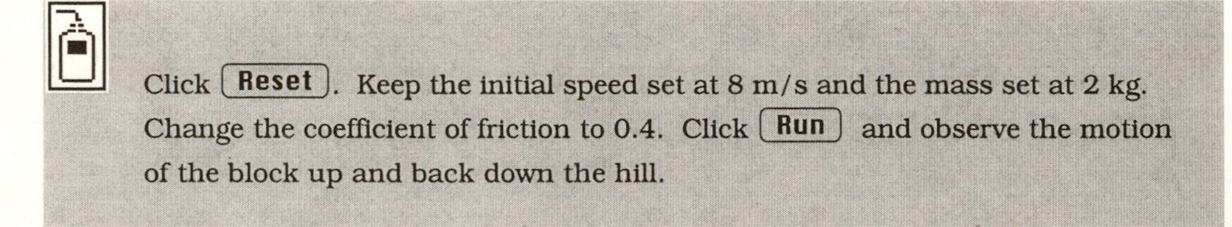

Click **Reset**. Keep the initial speed set at 8 m/s and the mass set at 2 kg. Change the coefficient of friction to 0.4. Click **Run** and observe the motion of the block up and back down the hill.

What differences (if any) are there between the graphs shown in the simulation and your predictions?

Use the simulation data to fill in the table below. Round the numbers in the second row to whole numbers.

Inclined Plane with Friction = 0.4

initial speed	distance - up	FF - up	work - up	FF - down	work - down

net-work	KE - initial	KE - final	ΔKE	% KE - lost	mass

Which of the following were affected by changing the coefficient of friction? Why?

the distance traveled up the hill:

the force of friction:

the total work done by the force of friction:

the initial kinetic energy:

the fraction of kinetic energy lost:

Click **Reset**. Change the initial speed to 7 m/s and leave the mass set to 2 kg and the coefficient of friction set to 0.4. Click **Run** and observe the motion of the block up and back down the hill.

Use the simulation data to fill in the table below. Round the numbers in the second row to whole numbers.

Inclined Plane with Friction = 0.4

initial speed	distance - up	FF - up	work - up	FF - down	work - down

net-work	KE - initial	KE - final	ΔKE	% KE - lost	mass

Which of the following were affected by changing the initial speed? Why?

the distance traveled up the hill:

the force of friction:

the total work done by the force of friction:

the initial kinetic energy:

the fraction of kinetic energy lost:

Click **Reset**. Leave the initial speed of the block set to 7 m/s and the coefficient of friction set to 0.4. Change the mass to 1 kg. Click **Run** and observe the motion of the block up and back down the hill.

Use the simulation data to fill in the table below. Round the numbers in the second row to whole numbers.

Inclined Plane with Friction = 0.4

initial speed	distance - up	FF - up	work - up	FF - down	work - down

net-work	KE - initial	KE - final	ΔKE	% KE - lost	mass

Which of the following were affected by changing the mass? Why?

the distance traveled up the hill:

the force of friction:

the total work done by the force of friction:

the initial kinetic energy:

the fraction of kinetic energy lost:

Optional:

- Vary the initial speed of the block and record data for the distance traveled up the hill for a fixed coefficient of friction. Plot distance vs. initial speed using a graphing program. *Is the graph a straight line?* Try d vs. v^2.

- Set the initial speed of the block to 8 m/s. Vary the coefficient of friction and record data for the distance traveled up the hill. Plot distance vs. coefficient of friction using a graphing program. *What does this graph tell you?*

Self-Test Questions for Simulation 18

The questions below pertain to the two sets of graphs of kinetic and potential energy for two blocks (A and B) projected up an inclined plane with friction. The two blocks have the same mass.

BLOCK A

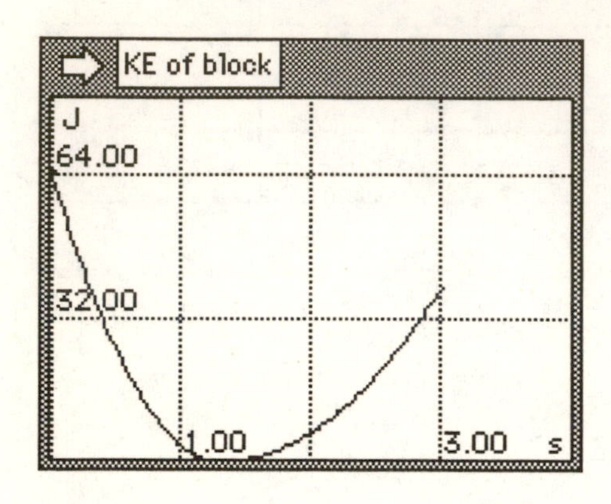

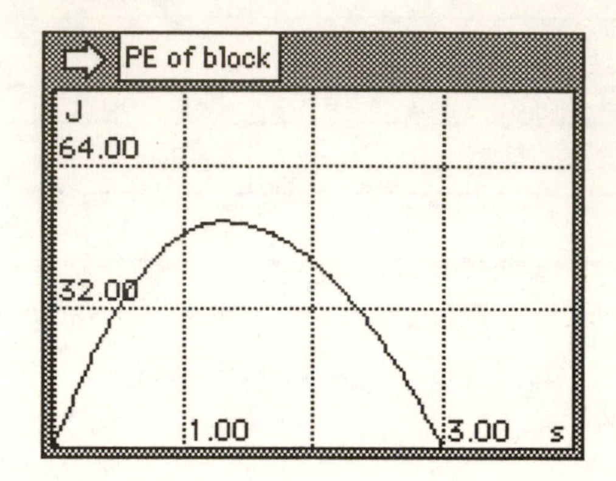

BLOCK B

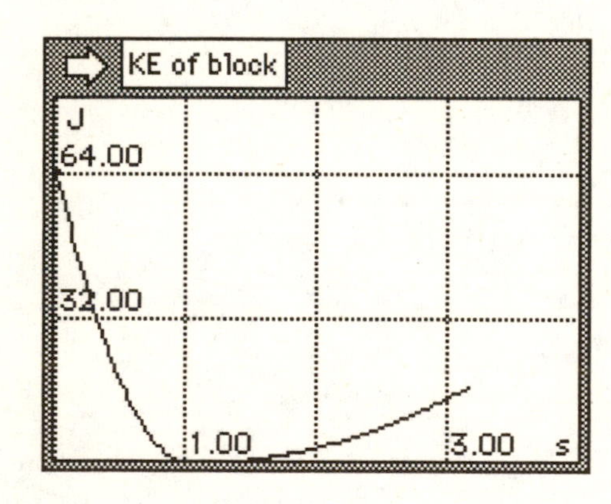

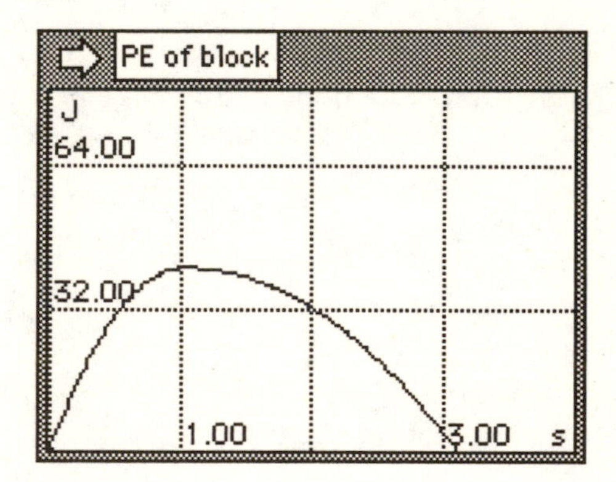

1. Which block lost more mechanical energy?
2. What percentage of the original energy is lost by block A?
3. Which block has more potential energy at t = 2 seconds?
4. Which block reaches its maximum height first?
5. Which block experiences the largest frictional force?

SIMULATION 19 A BLOCK SLIDING DOWN A HILL WITH FRICTION

Physics Review

As you saw in the previous two simulations, when the block was moving up the hill, the component of the force of gravity and the force of friction (*FR*) were in the same direction, down the hill. Therefore, when the block is moving up the hill, the net acceleration will always be directed down the hill, always slowing the block down, with an acceleration given by $a = g\sin\theta + \mu_k g\cos\theta$. By writing the acceleration this way, we have chosen the positive direction to be down the hill. When the block is moving down the hill, which is the situation you will explore in this simulation, the component of the force of gravity on the block and the force of friction are opposite in direction. The corresponding acceleration is given by $a = g\sin\theta - \mu_k g\cos\theta$ and it is possible for conditions to exist such that this acceleration is zero and the block will move at a constant speed. This will happen when $\mu_k = \tan\theta$. So for a given angle of inclination, the block will slide at a constant speed when the coefficient of friction is equal to the tangent of the angle of inclination. The other possibilities are that the coefficient of friction is less than the tangent of the angle, in which case the block will speed up as it moves down the hill, or that the coefficient of friction is greater than the tangent of the angle, in which case the block will slow down as it moves down the hill. Note that all of this is independent of the mass of the block.

Formulas you should be familiar with:

$FF = \mu_k FN$ $FN = mg\cos\theta$

Reference topics: friction, inclined planes

Simulation Details

The block, which has a mass of 2 kg, sits initially on a hill inclined at 21°. The only parameters that you will be able to adjust are the initial speed of the block and the coefficient of kinetic friction between the block and the hill. You have output meters that measure displacement (from the initial position), speed, and acceleration of the block. In this simulation the positive direction is defined as down the hill. The simulation will automatically stop when the block is near the bottom of the hill.

Calculate the coefficient of friction needed for the block to slide down the hill at a constant speed. You will need four digits after the decimal point to get an accurate result.

coefficient of friction = ____________

Set the initial speed of the block to 4 m/s and the coefficient of friction to the value that you just calculated. You will need to enter this number into the box. Click Run.

Is the acceleration of the block approximately zero?

Sketch the graphs of displacement and of speed of the block in the space below.

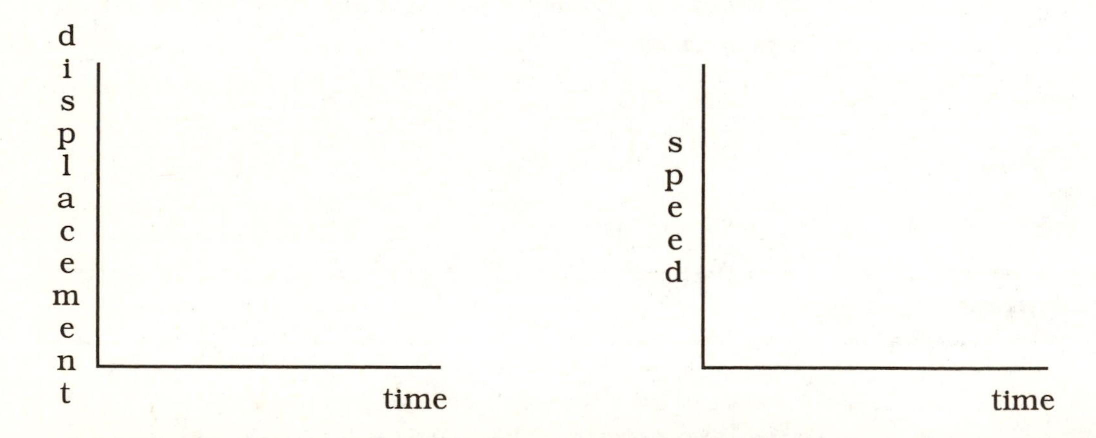

Click Reset. Set the coefficient of friction to a value equal to about half of the previous value. Do not run yet.

Predict what the displacement and speed graphs will be and sketch them below.

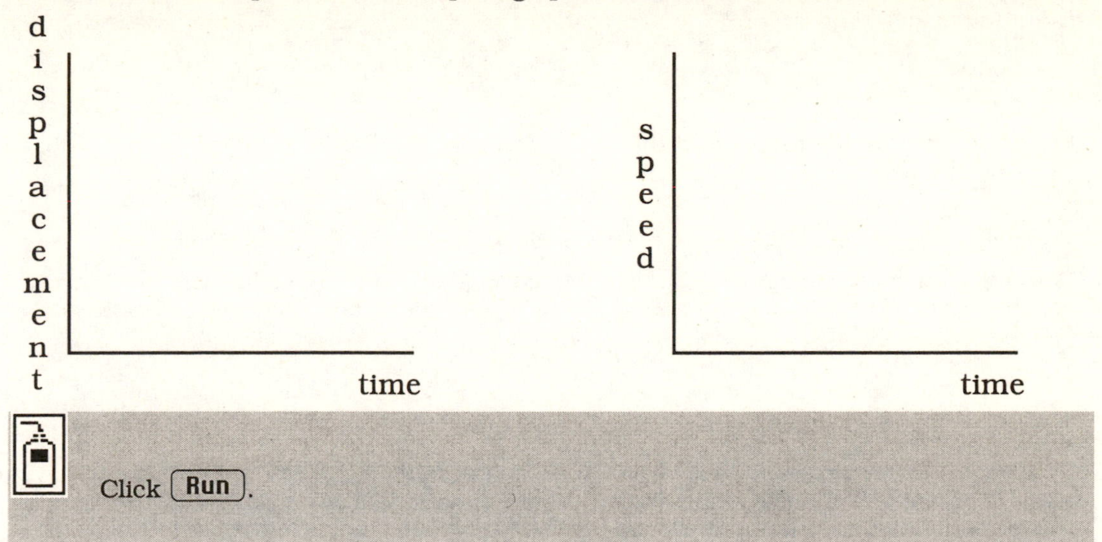

Click Run.

From the simulation data, record the value of the acceleration of the block.

acceleration = ______

Compare your predicted graphs of displacement and speed to those in the simulation. *Are they similar?*

Click Reset. Set the coefficient of friction to 0.5. Do not run yet.

Predict what the displacement and speed graphs will look like and sketch them below.

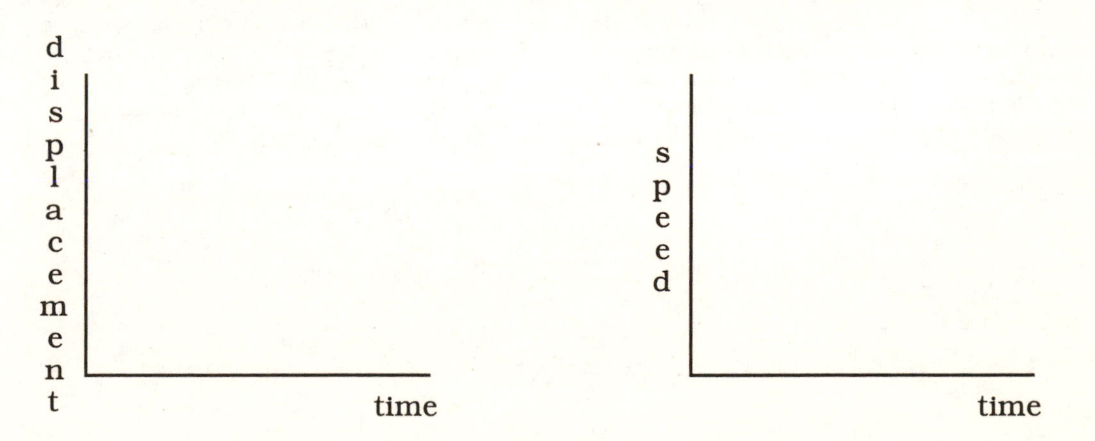

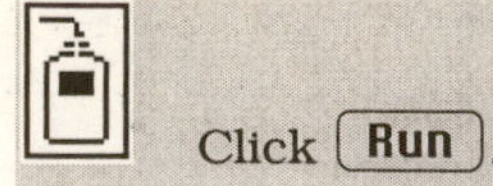

From the simulation data, record the value of the acceleration.

acceleration = ______

Compare your predicted graphs of displacement and speed to those in the simulation. *Are they similar?*

By running with different values of the coefficient of friction, try to duplicate the following graph of the speed of a block sliding down the hill. Just to make things a little easier for you, I will tell you that the only values for the coefficient of friction that were used to create this data were 0.0, 0.2, 0.4, and 0.6.

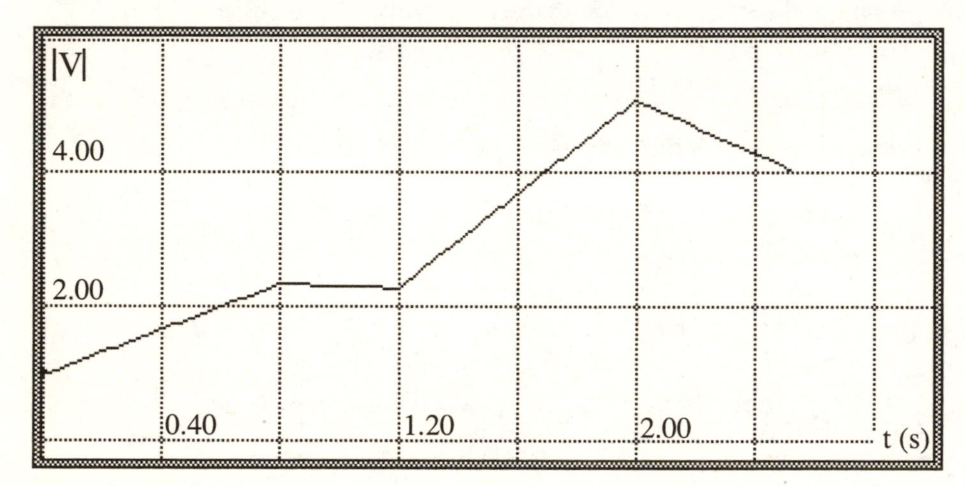

Fill in what you think the values are for the coefficient of friction and the initial speed. Then run with those values and draw your resulting graph of speed vs. time on the following page. Remember, you can stop, use the [O_O] to find a given frame, change the coefficient of friction, and then run from there. You must use the slider to change the coefficient of friction in this case. DO NOT TYPE A NUMBER INTO THE BOX.

initial speed = ______

$t < 0.8$	μ_k = ________
$0.8 < t < 1.2$	μ_k = ________
$1.2 < t < 2.0$	μ_k = ________
$t > 2.0$	μ_k = ________

Your Graph

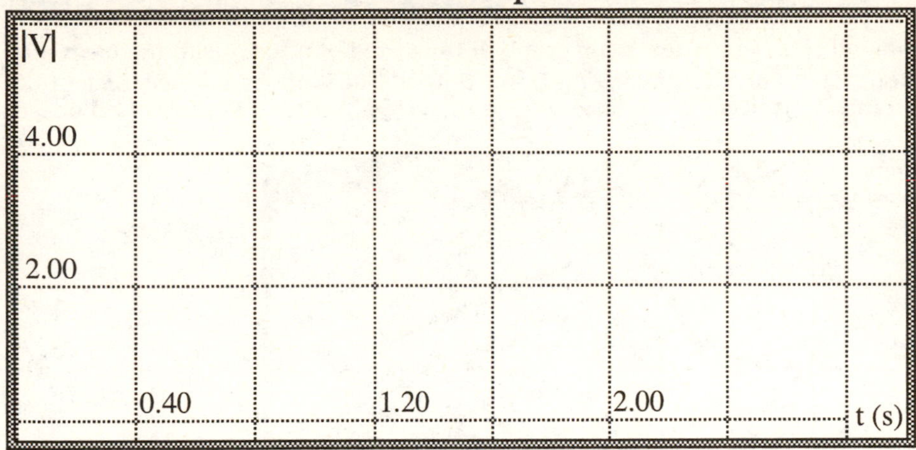

Compare your graph to the graph on the previous page. If they agree, skip to the self-test questions on the next page. If they do not, consider the following hint and try again.

Hint

The slope of the speed curve is related to the acceleration of the block. If the slope is positive (increasing), then the acceleration is positive, too. What value of coefficient of friction would give the greatest acceleration and hence the greatest slope? (0.0, 0.2, 0.4, or 0.6) What value would give a negative slope? (0.0, 0.2, 0.4, or 0.6) What value would give a near constant slope? (0.0, 0.2, 0.4, or 0.6)

$t < 0.8$	μ_k =________
$0.8 < t < 1.2$	μ_k =________
$1.2 < t < 2.0$	μ_k =________
$t > 2.0$	μ_k =________

|V|
4.00
2.00
0.40 1.20 2.00 t (s)

Self-Test Questions for Simulation 19

These questions all refer to the following graph of the speed of a block sliding down a hill inclined at about 21°. The simulation was paused after 1 second and the coefficient of friction was changed. It was also paused and changed again after 2 seconds had passed.

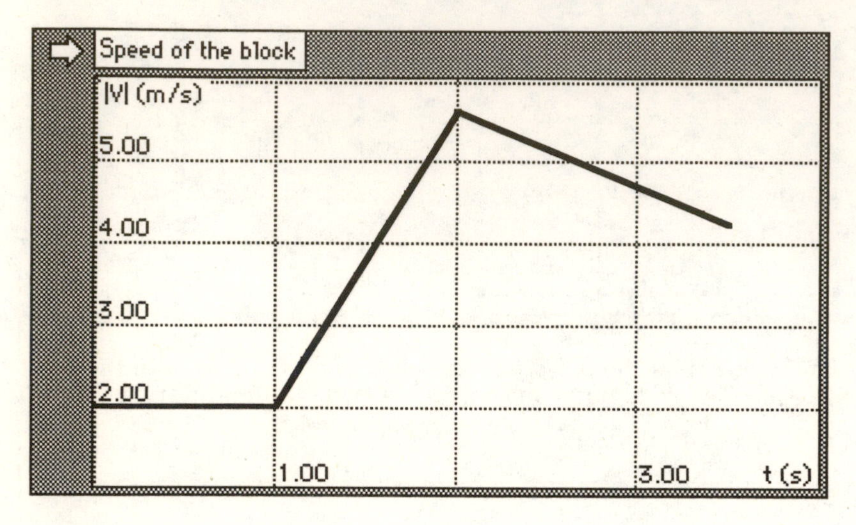

1. During what time interval is the coefficient of friction the smallest?
2. During what time interval is the coefficient of friction the largest?
3. Is there a time interval when the net force on the block is zero? If so, when?
4. During what time interval (if any) is the coefficient of friction equal to the tangent of the angle of inclination of the hill?
5. During what time interval (if any) is the coefficient of friction equal to zero?

SIMULATION 20 PULLING THREE CRATES

Physics Review

Newton's second law tells us that the net force on a body is equal to the mass of that body times the acceleration of that body. This is straightforward when a situation involves only one object, but one must be very careful in deciding what forces act on what objects when there are more than one. In this simulation we will be studying a system of three crates connected by ropes. One of the crates will be pulled on by an external force. Since the ropes have negligible mass, there are only three objects to consider. You can apply Newton's second law to each object separately and then solve the resulting set of equations for the tensions in the ropes and the acceleration of the crates (one value as we assume that the ropes do not go slack).

Formulas you should be familiar with:

$F = ma$ $\quad\quad$ $FF = \mu_k FN$

Reference topics: Newton's laws, friction

Simulation Details

In this simulation there are three crates resting on a horizontal floor. The first crate (the one on the right) is pulled by a horizontal force of adjustable magnitude. The direction of the force is fixed at 0° (with respect to the x direction). The crates are attached to each other by ropes. You can adjust the mass of each crate individually, and you can adjust the coefficient of kinetic friction between the floor and the crates. Output meters will measure the tension in each rope as well as the acceleration of the system. This simulation will stop when the first crate is near the end of the floor.

Let m_1, m_2, and m_3 represent the masses of the three crates and T_1 and T_2 the tensions as shown in the figure below. F is the applied external force.

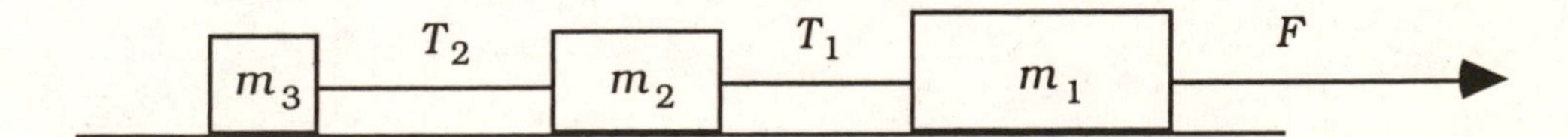

Assume a coefficient of friction, μ_k, and draw free body diagrams on each of the crates, showing all forces.

m_3 m_2 m_1

Apply Newton's second law ($F = ma$), in the direction of motion, to each body and write down the three resulting equations.

$m_1 a =$

$m_2 a =$

$m_3 a =$

Explanation

If you add these three equations (a common method for solving three equations with three unknowns), you should be able to solve for the acceleration and the tension given any values for the masses and the external force F. If you manipulate them, you should be able to get the following expression for the acceleration:

$$a = \frac{F}{M} - \mu_k g \qquad \text{where } M = m_1 + m_2 + m_3$$

Assume an external force of 20 N, m_1 = 5 kg, m_2 = 4 kg and m_3 = 1 kg. Calculate the acceleration of the crates and the tension in the ropes if there is no friction.

$a =$ ________

$T_1 =$ ________

$T_2 =$ ________

Set the coefficient of friction to 0.0 and the external force to 20 N. Set the mass of crate 1 to 5 kg, the mass of crate 2 to 4 kg, and the mass of crate 3 to1 kg. Click **Run** and observe for a short time.

From the simulation data, record or calculate the following:

a = ________

T_1 = ________

T_2 = ________

Do these values agree with the values that you calculated on the previous page?

Click **Reset**. Set the force to 40 N. Click **Run**.

From the simulation data, record or calculate the following:

a = ________

T_1 = ________

T_2 = ________

What changed when you doubled the force? Why?

Was anything unaffected by doubling the force?

Click **Reset**. Leave the force set to 40 N and the masses set to 5 kg, 4 kg, and 1 kg. Set the coefficient of friction to 0.2. Do not run yet.

Which of the following will change and how?

acceleration:

T_1:

T_2:

Click **Run** and observe the motion.

From the simulation data, record or calculate the following:

a = ________

T_1 = ________

T_2 = ________

Are you surprised that the tensions did not change?

Explanation

Notice that the acceleration decreased but the tension in the ropes was unaffected by a change in the coefficient of friction. If you go back to the three equations at the top of page 114, you can solve for the tensions to get

$$T_1 = \frac{(m_2 + m_3)}{M}F \quad \text{and} \quad T_2 = \frac{m_3}{M}F \quad \text{where } M = m_1 + m_2 + m_3$$

These expressions do not depend on the coefficient of friction, so the two different rope tensions will be the same for different coefficients of friction.

Assume that the total mass of the crates is 10 kg and the external force pulling on crate 1 is still 40 N. Calculate the acceleration of the crates if the coefficient of friction is 0.2.

a = ________

Click **Reset**. Set the force to 40 N and the coefficient of friction to 0.2. By varying the masses of the crates, try to achieve the following cases. In any try you must keep the total mass set at 10 kg. Record the mass values used to achieve each case and the resulting tensions and acceleration in the table below. Note that not all cases may be possible. Use different values for CASE 4 and CASE 5.

Investigating Different Mass Combinations
Total Mass Is 10 kg

	m_1	m_2	m_3	T_1	T_2	a
CASE 1: $T_1 > T_2$						
CASE 2: $T_2 = T_1$						
CASE 3: $T_2 > T_1$						
CASE 4: $T_1 = 2T_2$						
CASE 5: $T_1 = 2T_2$						

Explanation

If you look back at the data you just recorded, you should see that:

1) There is no way that the tensions can be equal (unless the middle mass is zero, which is not an option in this simulation), and, in fact, the tension in the first rope is always greater than the tension in the second rope.
2) There are several combinations that will give $T_1 = 2T_2$.
3) The acceleration is the same in all possible cases as it depends on the total mass and not the individual values.

You will need some data for the next simulation. Set the force to 40 N. Set the mass of crate 1 to 1 kg, crate 2 to 2 kg, and crate 3 to 1 kg. Set the coefficient of friction to 0.0 and click Run. Fill the simulation data into the table on page 120. Then reset and run with a coefficient of 0.1 and then with 0.2. Fill that data into the table as well.

Self-Test Questions for Simulation 20

The following statements refer to this set of crates being pulled on a horizontal surface with friction. True or false?

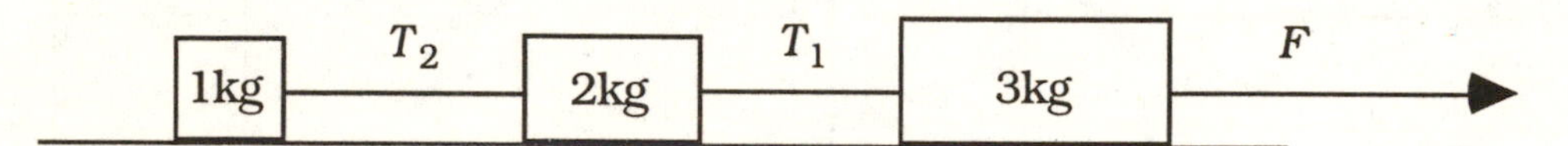

1. The tension T_1 can never be greater than F.
2. The tension T_2 can never be greater than T_1.
3. If the 2 kg and 1 kg crates are exchanged, the acceleration of the system will decrease.
4. The tension in each rope depends on the coefficient of friction.
5. The tension in each rope depends on the force F.

SIMULATION 21 PULLING THREE CRATES UPHILL

Physics Review

We will again be looking at the case of three crates, only now the three crates will be pulled up a hill. This will affect two things. First of all, since the normal force has changed, the force of friction (on each mass) will no longer be equal to $\mu_k mg$ but will be $\mu_k mg\cos\theta$. Since $\cos\theta$ is less than 1, this frictional force will be smaller than it was on the level surface (for any given coefficient of friction). This might lead one to conclude that the acceleration of the blocks would be larger; however, there is now a force equal to $mg\sin\theta$ acting down the inclined plane. This force will act to decrease the acceleration of the blocks. These two factors compete, and which one "wins" depends on the inclination of the hill (see the figure below, which shows the forces on the first crate). In the last simulation you saw that friction only affected the acceleration of the system and not the tensions in the ropes. In this simulation you will investigate how acceleration and tensions change (or do not change) when the same blocks are pulled up an inclined plane. As well, you will investigate the relationship between the work done by the external force, the work done by the frictional force, and the changes in mechanical energy (the extended work/energy theorem equation shown below).

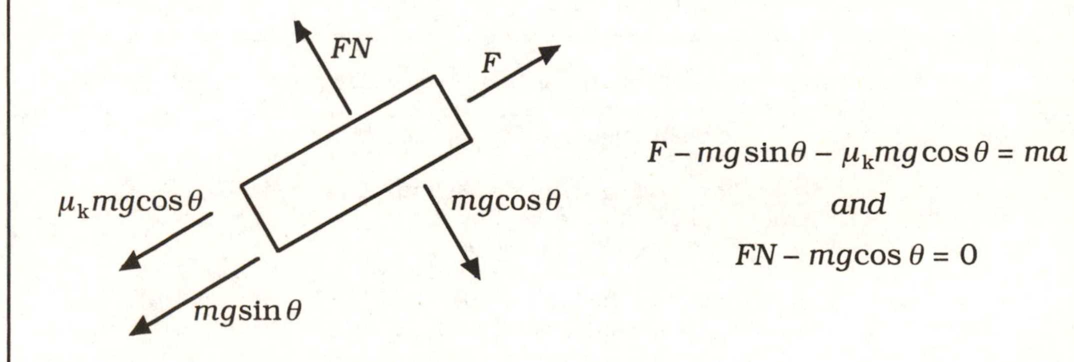

Formulas you should be familiar with:

$F = ma$ $\qquad FF = \mu_k FN$ $\qquad W_{net} = \Delta KE + \Delta PE$

Reference topics: Newton's laws, friction, work/energy theorem

Simulation Details

In this simulation the three crates will now be pulled up a hill inclined at 30°, and you will not be able to adjust that. The masses of the blocks, the coefficient of friction, and the external force, however, may all be varied. The simulation will pause when the blocks have been pulled a distance of approximately one meter. You cannot proceed with this simulation unless you have completed the previous one, since data from that simulation is needed for comparison to data in this one. In fact, it should already be in the table below. If it is not, you will have to go back and open simulation 20 to get that data.

On a Level Surface (F = 40 N)

coefficient of friction	T_1	T_2	a
0.0			
0.1			
0.2			

Now you will predict how things will change if these blocks are pulled uphill. Assume that the same force of 40 N is exerted on the first crate and that the masses are still 1 kg, 2 kg, and 1 kg respectively. In what follows, circle the correct word.

- For the case with no friction, and the crates pulled uphill, the magnitude of the acceleration of the crates will be *(greater than, less than, the same as)* the magnitude of the acceleration in row 1 of the above table.

- For the case with no friction, and the crates pulled uphill, the tension in the ropes will be *(greater than, less than, the same as)* the tension in the ropes in row 1 of the table.

- For the case with a coefficient of friction of 0.2, and the crates pulled uphill, the magnitude of the acceleration of the crates will be *(greater than, less than, the same as)* the magnitude of the acceleration in row 3 of the above table.

- For the case with a coefficient of friction of 0.2, and the crates pulled uphill, the tension in the ropes will be *(greater than, less than, the same as)* the tension in the ropes in row 3 of the table.

Set the force to 40 N. Set the mass of crate 2 to 2 kg and the others to 1 kg. Click **Run** and observe until the simulation pauses. Do this three times with a different coefficient of friction each time and fill the data into the table below. Reset each time.

On an Incline (F = 40 N)

coefficient of friction	T_1	T_2	a
0.0			
0.1			
0.2			

Compare these values to those in the table on the previous page.

Do you understand why the acceleration in each of these three cases is less than the acceleration in each corresponding case on the level surface? Explain.

Do you understand why the tensions have not changed in any of the three cases? Explain.

Assume the same force of 40 N, a coefficient of friction of 0.1 and that the first and second crates are switched. Predict whether the acceleration or tensions will change and how. In what follows, circle the correct word.

- The magnitude of the acceleration of the crates will be *(greater, less, the same).*
- The tension in the rope between the first two crates will be *(greater, less, the same).*
- The tension in the rope between the last two crates will be *(greater, less, the same).*

Set the force to 40 N and the coefficient of friction to 0.1. Set the mass of crate 1 to 2 kg and the masses of the other two crates to 1 kg. Click **Run** and observe the motion.

From the simulation data, record or calculate the following:

a = ________

T_1 = ________

T_2 = ________

Do you understand why the acceleration did not change when the two crates were switched? Explain.

Do you understand why only one tension changed when the two crates were switched? Explain.

From the simulation data, record or calculate the following:

the distance traveled by the crates = ________

the work done by the force of friction = ________

the work done by the external force = ________

the change in kinetic energy of the crates = ________

the change in potential energy of the crates = ________

Is the work/energy theorem valid? Check it using the data you just recorded.

Click **Reset**. Change the force to 50 N and leave all other settings alone.
Click **Run** and observe the motion.

From the simulation data, record or calculate the following:

a = __________

T_1 = __________

T_2 = __________

Do you understand why the acceleration has changed? Explain.

Do you understand why the tensions have changed? Explain.

From the simulation data, record or calculate the following:

the distance traveled by the crates = ______________

the work done by the force of friction = __________________

the work done by the external force = __________________

the change in kinetic energy of the crates = ______________

the change in potential energy of the crates = ______________

Compare these values to the values when the external force was only 40 N (page 122). Note that values that are very close can be considered to be the same as there are rounding errors.

Do you understand why the work done by the force of friction is the same in both cases? Explain.

Do you understand why the work done by the external force is different in each case? Explain.

Do you understand why the change in kinetic energy is greater for one case, yet the change in potential energy is the same in either case? Explain.

Self-Test Questions for Simulation 21

Consider three crates being pulled up a frictionless hill by a constant force of 10 N. The mass of the first crate is 3 kg and the others are 1 kg each. Which of the following statements are true and which are false?

1. The acceleration of the crates is less than 2 m/s^2.

2. Switching the first and second crate will change the acceleration.

3. Switching the first and second crates will change the tensions in both of the ropes between the three crates.

4. The change in kinetic energy is the same for each crate.

5. The change in potential energy is the same for each crate.

SIMULATION 22 A TWO-CAR COLLISION IN ONE DIMENSION

Physics Review

Two objects collide head-on in one dimension. This idealized collision where the direction of motion of each object is confined to the direction defined by the initial velocities is called a one-dimensional collision. All collisions conserve momentum ($\vec{p}$); that is to say, the total momentum of the system before the collision must be equal to the total momentum of the system after the collision. Typically, one can calculate the final velocities of each object, knowing the initial velocities and whether the collision is elastic (the case where kinetic energy is also conserved) or completely inelastic (the case where the objects stick together). An example of a collision that is almost elastic is one where two billiard balls collide on a pool table; an example of a completely inelastic collision is a piece of clay colliding with a block of wood. In reality most collisions are of neither type, and predicting final velocities can be difficult to do without a computer. The parameter that defines a collision is called the coefficient of restitution, which has a value of 1 for an elastic collision and a value of 0 for a completely inelastic collision.

Formulas you should be familiar with:

$$\vec{p} = m\vec{v} \qquad KE = \tfrac{1}{2}mv^2$$

Reference topics: collisions, elastic collisions, inelastic collisions, momentum

Simulation Details

The car that is initially on the left will be referred to as the LEFT CAR, and the car that is initially on the right will be referred to as the RIGHT CAR. You can adjust the initial velocity of each car separately. For this simulation the masses of the cars are equal and you will not be able to adjust them. Output devices will allow you to measure velocity graphs and the kinetic energy of each car. By varying the "collision elasticity" (effectively the coefficient of restitution), you will be able to investigate elastic collisions, completely inelastic collisions, and those in between.

Set the collision elasticity to completely elastic on the slider. Then set the velocity of the left car to some positive value and the velocity of the right car to a different negative value.

Record the values of initial velocity.

LEFT CAR initial velocity = ________

RIGHT CAR initial velocity = ________

Click Run and observe the motion of the cars.

Sketch the velocity graph of each car in the space below and identify the point on each graph where a collision occurred.

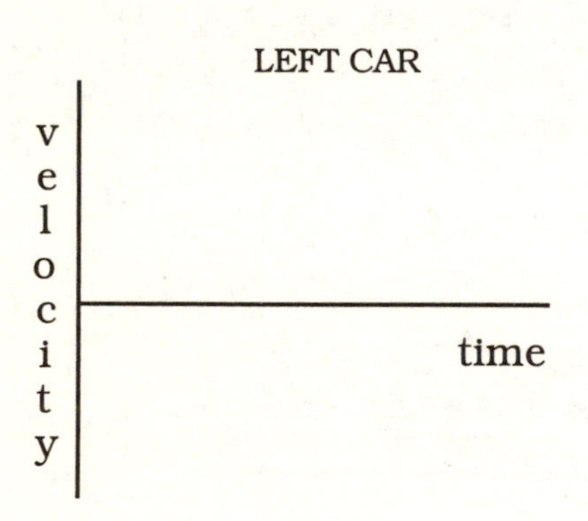

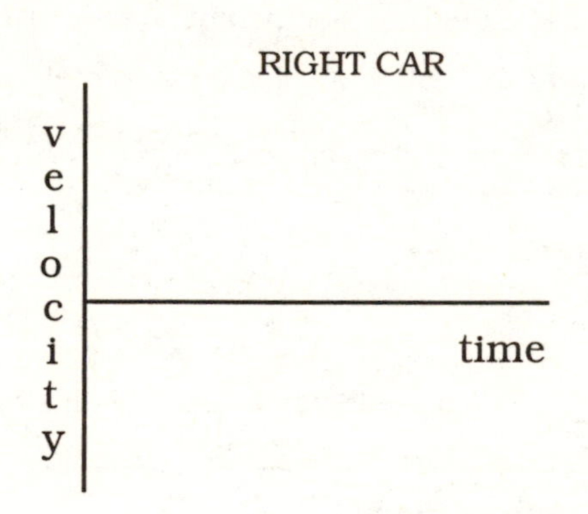

What is the final velocity of each car?

LEFT CAR final velocity = ________

RIGHT CAR final velocity = ________

From the simulation data, record or calculate the following (round to whole numbers):

the mass of each car = ____________

the initial momentum of the system = ____________

initial kinetic energy of the system = ____________

final momentum of the system = __________

final kinetic energy of the system = ________

percentage of total momentum lost = ____________

percentage of total kinetic energy lost = ____________

Click Reset. Do not change the initial velocities of the cars. Change the collision elasticity to completely inelastic. Do not run yet.

Calculate what the final velocity of each car will be and determine the direction they will be moving in after the collision.

direction of motion = ____________

final velocity = ____________

Predict what the velocity graph of each car will look like and draw them below.

LEFT CAR

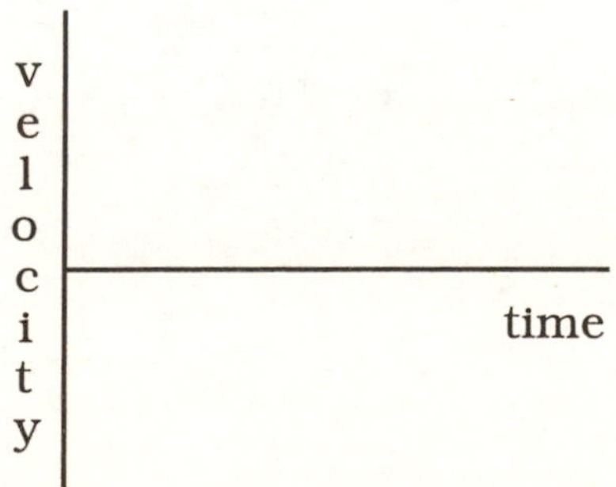

RIGHT CAR

velocity

time

Click [Run] and observe the motion.

What differences (if any) are there between the graphs you predicted on the previous page and those shown in the simulation?

How does your calculation of the final velocity compare to the value shown in the simulation?

From the simulation data, record or calculate the following (round to whole numbers):

the initial momentum of the system = ____________

initial kinetic energy of the system = ____________

final momentum of the system = __________

final kinetic energy of the system = ________

percentage of total momentum lost = ____________

percentage of initial kinetic energy lost = ____________

Click [Reset]. Keep the initial velocity of the cars the same. Set the collision elasticity to a value somewhere in the middle. Click [Run] and observe the motion.

From the simulation data, record or calculate the following (round to whole numbers):

the initial momentum of the system = ___________

initial kinetic energy of the system = ___________

final momentum of the system = ___________

final kinetic energy of the system = ___________

percentage of total momentum lost = ___________

percentage of initial kinetic energy lost = ___________

Which type of collision resulted in the greatest percentage of kinetic energy lost?

Which type of collision resulted in the greatest amount of momentum lost?

Click **Reset**. Set the initial velocity of the left car to 6 m/s and the initial velocity of the right car to -6 m/s. Run for each case listed in the table.

Type of Collision	Left Car - Final Velocity	Right Car - Final Velocity
elastic		
inelastic		
in between		

What conclusions can you draw from the data in your table? Remember, these are cars with equal mass; whatever you conclude may not be valid for cars with different masses.

Click **Reset**. Set the initial velocity of the left car to –1 m/s and the initial velocity of the right car to – 9 m/s. Do not run yet.

Do you think that the cars will collide? Why or why not?

Click **Run** to test your prediction.

Did they collide? Approximately when did they collide?

Optional: View the collision from the reference frame of the drivers of the left and right cars. *Are they what you expected?* Try different initial velocities.

Self-Test Questions for Simulation 22

True or false?

1. All collisions conserve momentum.
2. All collisions conserve kinetic energy.
3. A completely inelastic collision results in a maximum amount of kinetic energy lost.
4. For any perfectly elastic collision between two objects of equal mass, they will always exchange velocities.
5. Completely inelastic collisions (where the objects stick together) typically involve energy losses of only about 10%.

SIMULATION 23 A BALL THROWN UP INTO THE AIR (ENERGY)

Physics Review

We have investigated projectile motion under several different conditions: free fall motion that is only vertical, planar motion with and without air resistance. When we did these, we studied the relationships between position, velocity, and acceleration primarily using the kinematic equations for motion with constant acceleration (except in the case with air resistance). It is often easier (and sometimes more instructive) to study this type of motion by considering energy instead. The two types of energy of interest here are kinetic (*KE*) and gravitational potential (*PE*). The formulas for these are shown below, and you must remember that the choice of $y = 0$, and hence the point where potential energy is zero, can be different for different problems. You are free to choose the location of $y = 0$, but once chosen, all values must be measured with respect to that point. When the projectile is at a value of y that is positive, its potential energy will be positive, and when its y position is negative, its potential energy will be negative. What matters when dealing with gravitational potential energy is not its absolute value, but the change in its value during the motion. When there is no air resistance affecting the projectile, its total energy ($E = KE + PE$) will be constant. The force on the ball due to air resistance is called a dissipative force since it reduces (or dissipates) energy away from energy of motion.

Formulas you should be familiar with:

$$KE = \tfrac{1}{2}mv^2 \qquad PE = mgy$$

Reference topics: energy, kinetic energy, gravitational potential energy

Simulation Details

When you click on the Run button, the ball will be launched into the air at a speed and angle that you can adjust. The initial position of the ball is taken as the origin (0,0) at the edge of the cliff. In the first part the ball will be thrown straight up and you will study graphs of kinetic and potential energy. When the ball returns to its original position, the simulation will automatically pause. You should reset at that point. In the next part you will launch the ball at an angle so that it will rise, come down, and go over the cliff and fall into the canyon below. When the simulation pauses, you should click Run to see the ball fall into the canyon. In both cases you will be able to see the effects of air resistance on kinetic, potential, and total energy of the ball. "Zero" on the air resistance slider means that there is no air resistance. "Max" on the slider sets the air resistance force to have the maximum effect for this simulation.

Set the initial speed to 10 m/s, the mass to 1.25 kg and the angle to 90°. Check that the air resistance is set to zero. Click Run and observe the motion and the graphs of kinetic, potential, and total energy.

From the simulation data, record or calculate the following:

time of total flight = ________

maximum height attained = __________

In what follows, circle the correct word.

- The kinetic energy is (*increasing, decreasing*) when the ball is on the way up, and it is (*increasing, decreasing*) when the ball is on the way down.
- The potential energy is (*increasing, decreasing*) when the ball is on the way up, and it is (*increasing, decreasing*) when the ball is on the way down.

Sketch the graphs of kinetic and potential energy in the space below and indicate the following points by placing the corresponding letters on each graph:

launch = A
halfway up to the top = B
at the top = C
halfway down = D
back at the ground = E

KE

time

PE

time

Record the following simulation data into the table:

Ball Thrown Straight Up

	launch	halfway up	at the top	halfway down	at the edge
KE					
PE					
E					

Explanation

Notice that the potential energy and the kinetic energy are equal when the ball is located at a height that is half of its maximum height. *Do you think that this will be true regardless of the initial speed and mass of the ball?*

Predict what the graphs of kinetic, potential, and total energy will look like if you change the velocity to a larger value, say 12 m/s, and to a smaller velocity, say 8 m/s. Sketch them in the region below (use different colors or dotted lines). A reference time is shown that represents the time it takes for the ball to get to the top when the initial velocity is 10 m/s.

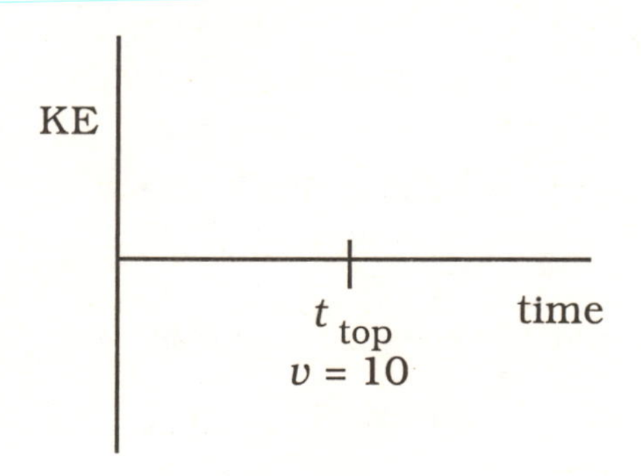

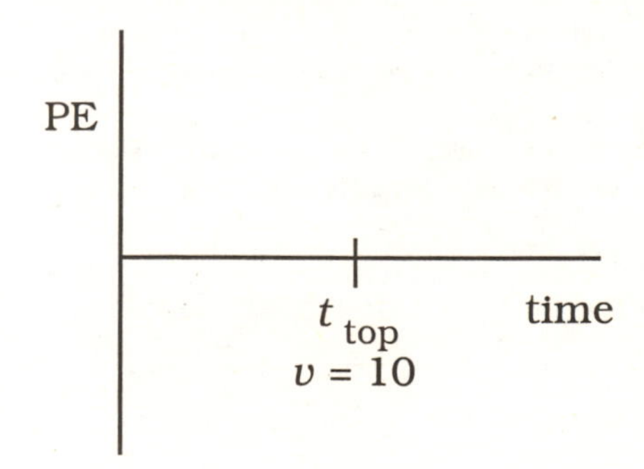

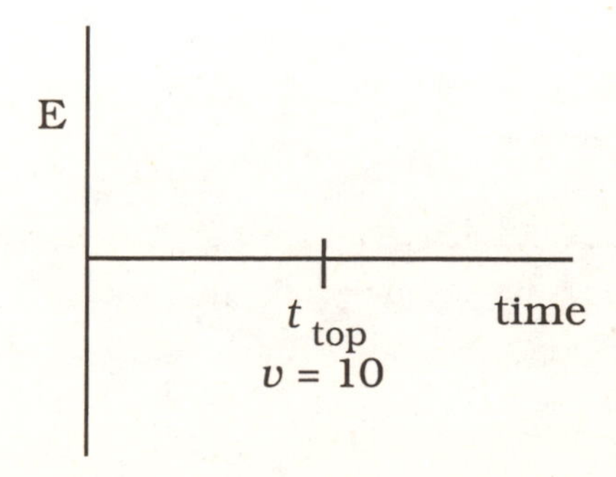

Click **Reset** and run with an initial speed of 8 m/s and then again with 12 m/s to test your predictions.

What happens to the kinetic energy graph as the initial speed is increased? Does it change shape? How?

What happens to the potential energy graph as the initial speed is increased? Does it change shape? How?

What happens to the total energy graph as the initial speed is increased? Does it change shape? How?

From the simulation data for an initial speed of 12 m/s, record the following:
(Do not include the last frame. Use the [O_O] to go back one frame and then fill in.)

maximum kinetic energy = ______________

minimum kinetic energy = ______________

maximum potential energy = ______________

minimum potential energy = ______________

time to the top = ______________

Click **Reset**. Set the initial speed to 12 m/s and decrease the mass of the ball to 1.0 kg. Click **Run** and observe the motion.

From the simulation data for an initial speed of 12 m/s record the following:
(Do not include the last frame. Use the [O O] to go back one frame and then fill in.)

maximum kinetic energy = ______________

minimum kinetic energy = ______________

maximum potential energy = ______________

minimum potential energy = ______________

time to the top = ______________

Does changing the mass affect the motion or the energy curves? How?

Sketch the graphs shown in the simulation for this case (m = 1 kg). Indicate some sort of scale by putting in values for t-top and final time. Also indicate the maximum values for *KE* and *PE* on the vertical axis.

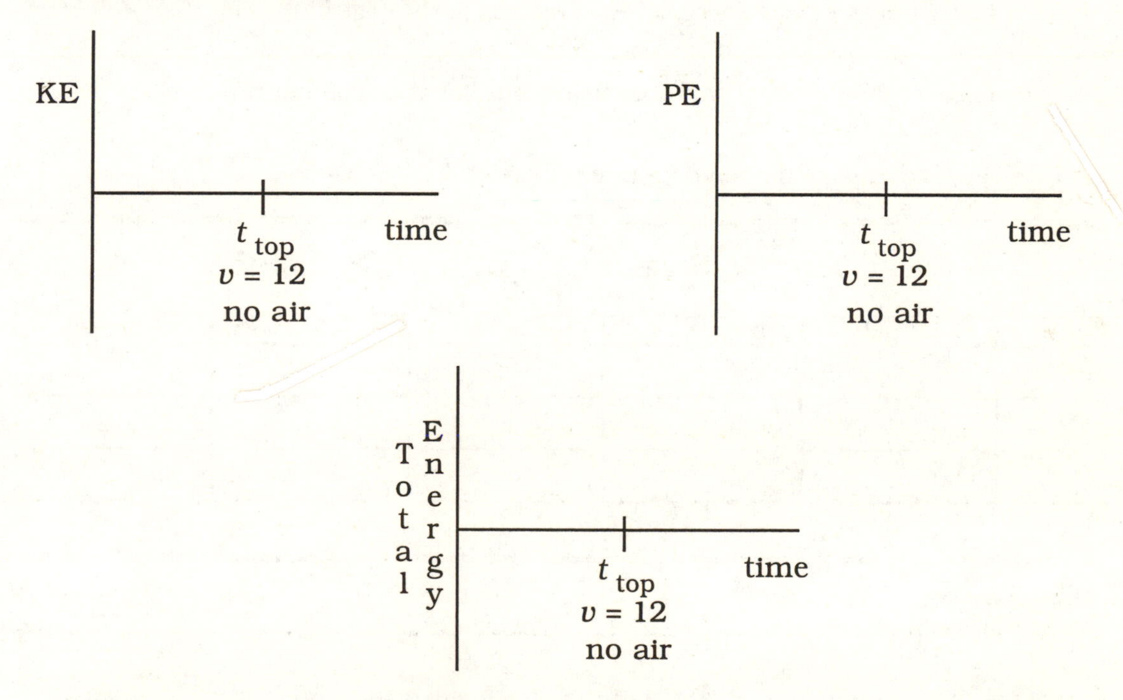

Predict what the graphs of kinetic energy, potential energy, and total energy will look like when air resistance is included (no change in anything else) and sketch them on the same graphs above that you used for the no air resistance curves (use a colored pencil or a dotted line).

Click [Reset]. Keep the same values of initial speed, direction and mass of the ball. Turn on air resistance by sliding to "max" on the air resistance slider. Click [Run] and observe until the pause.

These graphs are difficult to predict, so yours may not look exactly like the ones shown in the simulation. In general the important features that you should understand are:

- The kinetic energy reaches a value of zero (and the potential energy is a maximum) at a time that is slightly less than the time to the top with no air resistance.
- The maximum value of the kinetic energy is the initial value, which is the same.
- The maximum value of the potential energy (when the ball is at the top) is less than the initial kinetic energy because some of the energy is lost due to air resistance.
- Each energy curve is no longer symmetric with respect to up and down motion.
- The ball has less kinetic energy when it returns to the edge of the cliff than it did initially.
- The total mechanical energy is decreasing, but not at a constant rate.

Record the following simulation data into the table:

Ball Thrown Straight Up with Air Resistance

	launch	halfway up	at the top	halfway down	at the edge
KE					
PE					
E					

Explanation

Notice that the potential energy and the kinetic energy are no longer equal when the ball is located at a height that is half of its maximum height. The symmetry has been broken.

Now you will launch the ball at an angle of 30° with a speed of 10 m/s so that the ball will go over the cliff and into the canyon. This will be done with and without air resistance.

Click **Reset**. Set the initial velocity of the ball to 10 m/s, the ball's direction to 30°, the mass of the ball to 1 kg and air resistance to zero. Click **Run**. Click **Run** to go on from the pause and watch the ball fall into the canyon.

Sketch the graphs shown in the simulation in the space below. The time the ball is back near $y = 0$ is marked as well as the total initial energy of 50 J.

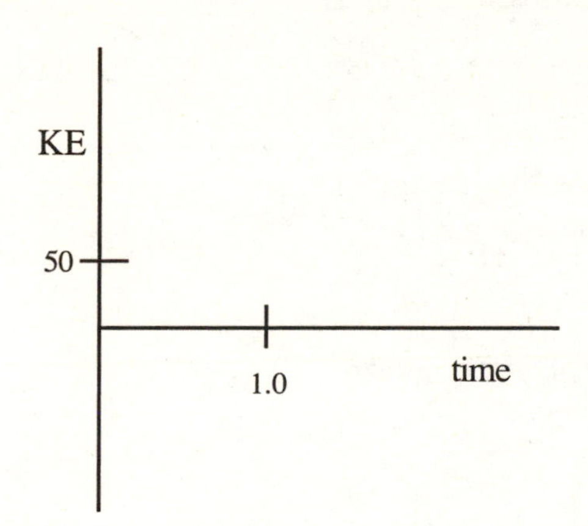

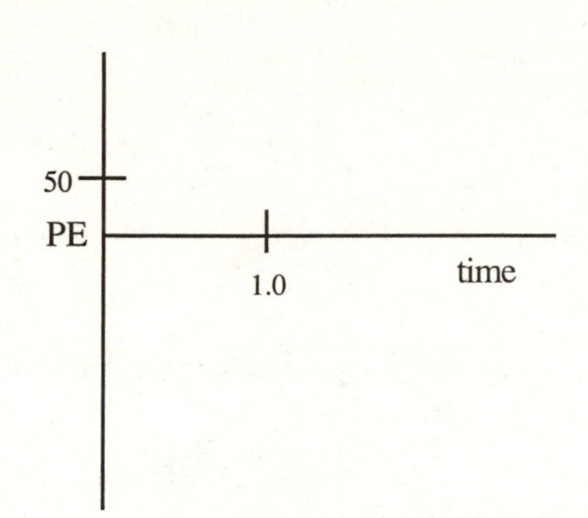

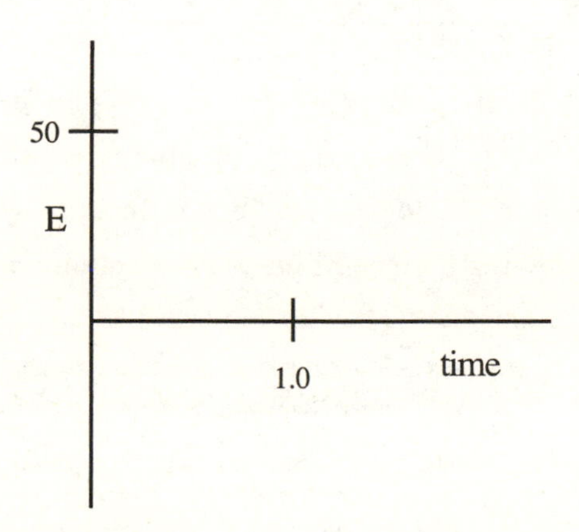

From the simulation data, record the following:

maximum kinetic energy = ________________

minimum kinetic energy = ________________

maximum potential energy = ________________

Click **Reset**. Keep everything set but change the air resistance to "max".
Click **Run**.

How are these graphs of energy different from the ones without air resistance that you sketched on the previous page?

From the simulation data, record the following:

maximum kinetic energy = ____________

minimum kinetic energy = ____________

maximum potential energy = ____________

Is the kinetic energy ever zero? Is the kinetic energy ever negative?

Is the potential energy ever zero? Is the potential energy ever negative?

Explanation

If you could keep running longer, eventually the ball would attain its terminal (constant) velocity, and the kinetic energy would approach a constant value as well. The potential energy would continue to get more and more negative, the graph approaching a straight line. Think about why this is so.

Self-Test Questions for Simulation 23

For a ball launched with a given speed, which statements are true and which are false?

1. The initial kinetic energy of the ball depends on the launch angle.
2. The potential energy is a maximum at the top of the flight.
3. If the ball is launched at 30°, with no air resistance, the kinetic energy at the top is zero.
4. The kinetic energy of the ball launched with no air resistance is always positive.
5. When a ball is launched at 30° with air resistance, its kinetic energy decreases, then increases, then approaches a constant value.

SIMULATION 24 A BALL THROWN UP THAT HITS THE CEILING

Physics Review

Physics problems do not always involve one concept, and in this simulation we will study a process that involves several. Two types of energy will be of interest here, kinetic (*KE*) and gravitational potential (*PE*) of a ball. The formulas for these are shown below, and you must remember that the zero for potential energy is arbitrary in any given situation as you are free to choose your reference point for where y = 0 and the potential energy is zero. The ball (thrown straight up in the air) will not be able to complete its motion as it will collide with a ceiling on its way upward. The energy lost in the collision with the ceiling can be varied.

Formulas you should be familiar with:

$KE = \frac{1}{2}mv^2$ $PE = mgy$

Reference topics: kinetic energy, gravitational potential energy, collisions

Simulation Details

The initial speed of the ball will be fixed at 10 m/s, and it will be thrown straight up. The initial position of the ball is y = 0.0 m, and hence its initial potential energy is zero. The mass of the ball will be adjustable with a slider. The ceiling position (defined with respect to the initial position of the ball) will be adjustable, as will the energy lost in the collision between the ball and the ceiling. You will investigate the effect of the collision on the time of flight of the ball, the final velocity of the ball, and the energy of the ball.

If unobstructed by the ceiling, calculate how high the ball will go.

maximum height of the ball = ____________

Set the ceiling position to some value that will assure that the ball does not hit it. Set the mass of the ball to 0.1 kg. Click **Run** and observe the motion.

Record the following from the simulation data:
(You will need to approximate time from the graph.)

time for the total flight = __________

initial velocity of the ball = ________

velocity of the ball when it gets back to its original position = __________

initial kinetic energy of the ball = _________

kinetic energy of the ball when it gets back to its original position = ________

What does it mean that the final velocity is negative?

Is the initial velocity equal to the final velocity?

Is the initial speed equal to the final speed?

Sketch the velocity graph in the space below. Mark the point where it turns around with an X.

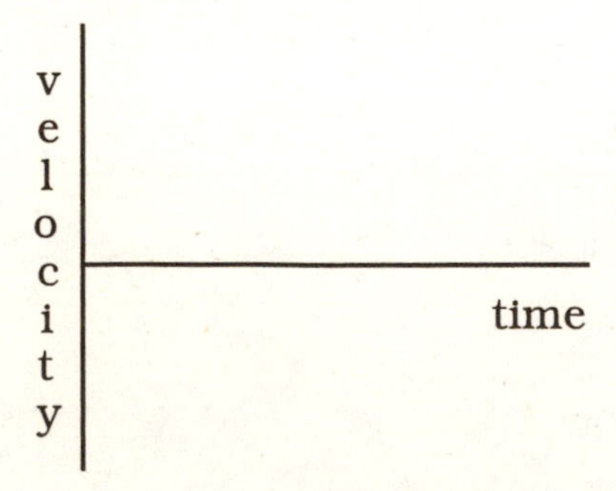

Now we will think about and examine a different situation. Suppose the ceiling was lower than the maximum height attained by the ball and the ball collides with the ceiling on the way up. Think about this a little bit, and then answer these questions.

Do you think that the time it takes the ball to get back to its original position will be greater, less, or the same? Why? What might it depend on?

Do you think that the speed of the ball when it is back to its original position will be greater, less, or the same as the initial speed? Why? What might it depend on?

Now let's investigate this situation.

Click **Reset**. Set the ceiling height to 3 m and the energy lost in the collision to "none". Click **Run** and observe the motion of the ball all the way until it returns to its original position.

From the simulation, record or calculate the following:

speed of the ball when it gets back to its original position = ____________________

kinetic energy of the ball when it gets back to its original position = ____________

time for the total flight = _______

Did you expect that the time would be less? Why?

Did you expect that the speed would be the same? Why?

Explanation

When the ball collides with the ceiling in a perfectly elastic collision, no energy is lost and the ball's velocity maintains its magnitude and changes direction. Had the ball been allowed to rise completely and then fall, when it got back to that point (the collision point), its velocity would be exactly the same. So as far as the final velocity goes, there is no change. The time, however, will be shorter as part of the trip has been omitted.

Sketch the velocity graph in the space below and mark the collision point with an X.

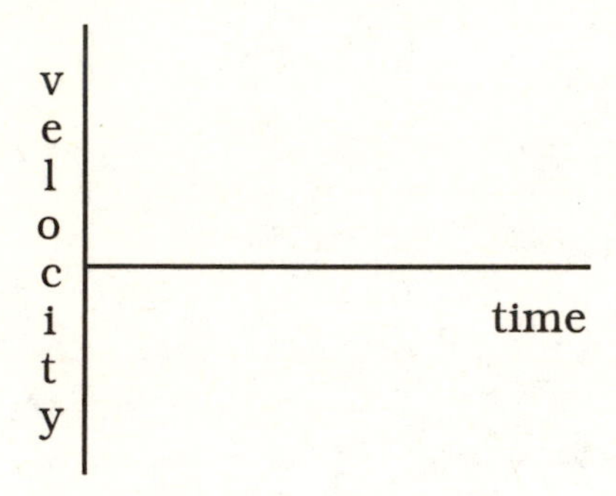

How is this graph different from the one in which there was no collision with the ceiling?

From the simulation data, record or calculate the following:

initial kinetic energy of the ball = __________

kinetic energy of the ball just prior to the collision with the ceiling = ____________

potential energy of the ball just prior to the collision with the ceiling = __________

kinetic energy of the ball just after the collision with the ceiling = ______________

potential energy of the ball just after the collision with the ceiling = ____________

kinetic energy of the ball when it gets back to its original position = _________

Calculate the percentage of kinetic energy lost in the collision:

% of kinetic energy lost in the collision = __________

Calculate the percentage of energy lost overall (i.e., from beginning to end):

% of energy lost overall = __________

Click **Reset**. Change the energy lost in the collision to "all" and keep the ceiling height set to 3 m and the mass set at 0.1 kg. Do not run yet.

In what follows, circle the correct word.

- The speed of the ball when it returns to the starting height will be (*equal to, greater than, less than*) the initial speed.
- The total time of the flight will be (*equal to, greater than, less than*) the total time in the previous case (elastic collision with the ceiling).

Predict what the velocity graph will look like now and sketch it in the space below.

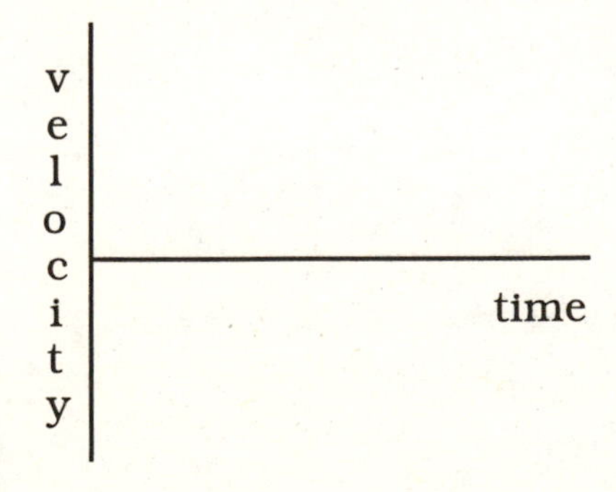

Click (Run) and observe the motion of the ball all the way until it returns to its original position.

From the simulation, record or calculate the following:

speed of the ball when it gets back to its original position = ____________

time for the total flight = _______

Did you predict that the speed of the ball would be less for this case? Why?

Did you expect that the time would be greater? Why?

Did your prediction of the velocity graph agree with what was shown in the simulation?

What is the slope of the velocity curve prior to the collision, and what is the slope of the velocity curve after the collision?

Will the slope of the velocity curve (except during the collision) be the same regardless of the collision elasticity? Why?

Explanation

When the ball collides with the ceiling and loses all of its kinetic energy, the ball's velocity is zero immediately after the collision. The ball will then fall freely from that point, but it starts out with a velocity of zero and therefore will never attain the same velocity at the end as it did in the previous case. For the same reasons, the time to fall from that point will be greater.

From the simulation data, record or calculate the following:

initial kinetic of the ball energy = ________

kinetic energy of the ball just prior to the collision with the ceiling = ________

potential energy of the ball just prior to the collision with the ceiling = ________

kinetic energy of the ball just after the collision with the ceiling = ________

potential energy of the ball just after the collision with the ceiling = ________

kinetic energy of the ball when it gets back to its original position = ________

Calculate the percentage of kinetic energy lost in the collision:

% of energy lost in the collision = ________

Calculate the percentage of energy lost overall (i.e., from beginning to end):

% of energy lost overall = ________

Optional: All these exercises can be done rather quickly if you have a graphing program on your computer or a graphing calculator.

- Keep the mass of the ball set to 0.1 kg and the energy lost in the collision set to "all" and vary the ceiling height. Investigate how the total energy lost ($E_{initial} - E_{final}$) is related to the height of the ceiling. Plot h vs. energy lost. You should get a straight line. *What is its slope?*

- Keep the mass of the ball set to 0.1 kg and the energy lost in the collision at an in-between value and vary the ceiling height. Investigate how the total energy lost ($E_{initial} - E_{final}$) is related to the height of the ceiling. Plot h vs. energy lost. You should get a straight line. *What is its slope?*

- Keep the ceiling height fixed at 3 m and the energy lost to somewhere in the middle and investigate what effect the changing the mass of the ball has on final speed and the percentage of initial total energy lost.

Self-Test Questions for Simulation 24

All of the questions refer to the following graphs of the velocity of a ball that was thrown straight up in the air. In both cases the ball made the same kind of collision (in terms of elasticity) with the ceiling, and in each case the mass of the ball is the same. Both graphs end when the ball returns to its original position.

A

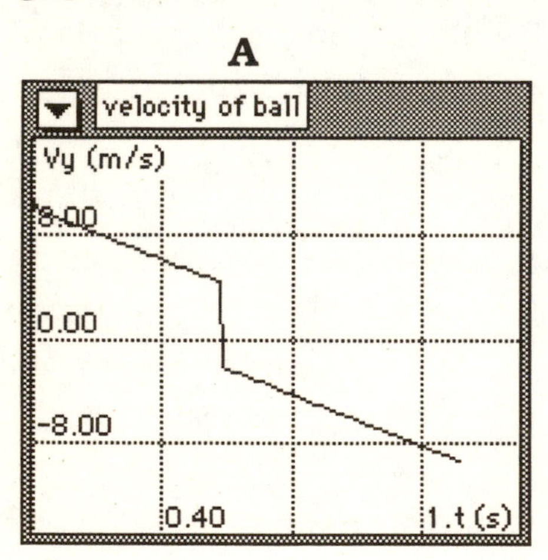

B

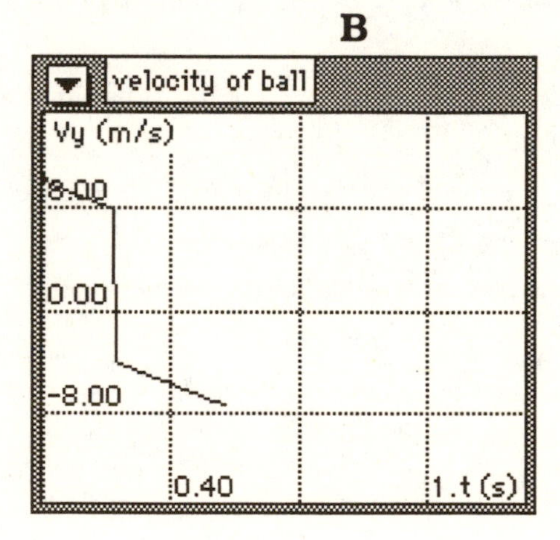

1. Is the initial kinetic energy of the ball in A greater than, less than, or equal to the initial kinetic energy in B?

2. Were the collisions (A and B) elastic?

3. Which graph represents the case with the lower ceiling?

4. Which graph represents the ball that lost more energy overall?

5. Which graph represents the ball with the smaller speed after the collision?

SIMULATION 25 UNIFORM CIRCULAR MOTION IN A HORIZONTAL PLANE

Physics Review

Many objects move in circular (or near circular) paths and so it is an important type of motion to understand. Circular motion is characterized by two parameters, the speed of the object (v) and the radius of its circular path (R). The object completes one cycle by traveling a distance of $2\pi R$ (the circumference of the circle) in a time T (the period). The frequency is measured in cycles/second (hertz) and is the reciprocal of the period.

Formulas you should be familiar with:

$$v = \frac{2\pi R}{T} \qquad f = \frac{1}{T}$$

Reference topics: circular motion, period, frequency

Simulation Details

A ball is moving in a horizontal circle on a smooth table. You are looking down on it from above. You will be able to adjust the radius of the circle as well as the mass and the velocity of the ball. The initial velocity direction will be either to the right (a positive number) or to the left (a negative number). As the simulation progresses, the velocity vector and its components will be shown on your computer screen. One graph will plot both the x and y components of the velocity.

Set the mass to 0.3 kg, the initial velocity to 8 m/s, and the length of the rod to 1.6 m. Do not run yet.

Calculate the period and frequency of the motion.

period = ____________

frequency = _________

Click **Run** and observe the motion for 3 seconds.

From the velocity graph, determine the frequency. Hint: Use the tape player to locate the point where t = 1 s. *How many cycles have occurred by that point?*

f = __________ cycles/sec

Sketch the graphs of the x and y components of velocity as shown in the simulation. Mark the following points (if they exist):

A -- one place where the two components of velocity are equal
B -- one place where the y component of the velocity is a maximum
C -- one place where the x component of the velocity is a maximum
D -- one point where both components are zero

V_x (m/s)
V_y (m/s)
0.00
-16.00
1.00 3.00 t (s)

Explanation

The graph for the x component of the velocity is a cosine curve and the graph of the y component of the velocity is a sine curve. The value on one curve is a maximum (or minimum) when the value on the other curve is zero. These two curves are said to be "out of phase by 90°" or "have a phase difference of 90°." We can write

$v_x = v_o \cos(2\pi ft)$ and $v_y = v_o \sin(2\pi ft)$

as the equations for the x and y components of the velocity. Note that these equations depend on the initial position of the ball. If it had started straight out to the right side of the pivot point with an initial velocity in the y direction, the x curve would be a sine and the y curve would be a cosine.

Now you will investigate the motion in more detail. We will look at four time regions separately. Use the tape player to locate points of interest.

During the first quarter of a cycle ($t < T/4$)

- The ball is moving upward and to the (*right, left*).
- The x component of the velocity of the ball starts out with a magnitude of _________ and (*increases, decreases*) to a value of _______.
- The y component of the velocity of the ball starts out with a value of _______ and (*increases, decreases*) to a magnitude of _______.
- The magnitude of the velocity of the ball (*remains constant, is changing*).
- Draw the velocity vector and its components at $t = 0.20$ sec.

During the second quarter of a cycle ($T/4 < t < T/2$)

- The ball is moving upward and to the (*right, left*).
- The x component of the velocity of the ball starts out with a value of _______ and (*increases, decreases*) to a magnitude of _______.
- The y component of the velocity of the ball starts out with a magnitude of _________ and (*increases, decreases*) to a value of _______.
- The speed of the ball (*remains constant, is changing*).
- Draw the velocity vector and its components at $t = 0.50$ sec.

During the third quarter of a cycle ($T/2 < t < 3T/4$)

- The ball is moving downward and to the (*right, left*).
- The x component of the velocity of the ball starts out with a magnitude of _________ and (*increases, decreases*) to a value of _______.
- The y component of the velocity of the ball starts out with a value of _______ and (*increases, decreases*) to a magnitude of _______.
- The speed of the ball (*remains constant, is changing*).
- Draw the velocity vector and its components at $t = 0.80$ sec.

During the last quarter of a cycle ($3T/4 < t < T$)

- The ball is moving downward and to the (*right, left*).
- The x component of the velocity of the ball starts out with a value of ________ and (*increases, decreases*) to a magnitude of ________.
- The y component of the velocity of the ball starts out with a magnitude of ________ and (*increases, decreases*) to a value of ________.
- The speed of the ball (*remains constant, is changing*).
- Draw the velocity vector and its components at t = 1.10 sec.

Click **Reset**. Change the initial velocity to –8 m/s and leave the mass set to 0.3 kg and the length of the rod set to 1.6 m. Do not run yet.

Calculate what the value of period and frequency will be now.

period = ____________

frequency = ____________

Click **Run** and observe the motion for 3 seconds.

From the simulation data, record or calculate the following:

period = ____________

frequency = ____________

How do these values compare with your calculations?

Sketch the graphs of the x and y components of velocity as shown in the simulation.

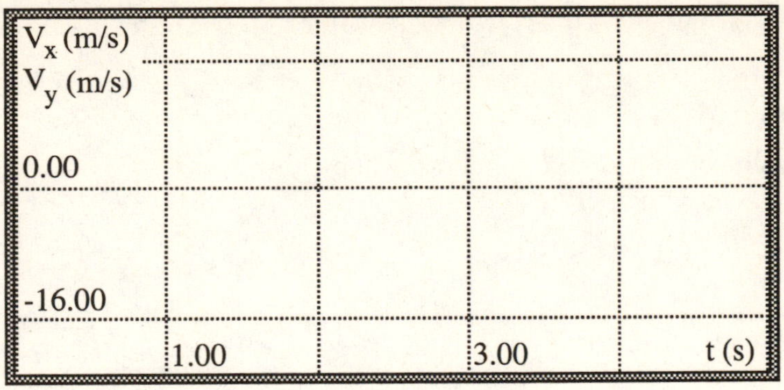

V_x (m/s)
V_y (m/s)

0.00

-16.00

1.00 3.00 t (s)

How are these curves different from the curves that you sketched on page 148, when the initial velocity was positive?

Which curve is represented by $-v_o \cos(2\pi ft)$*?*

Which direction is the ball moving in? Clockwise or counterclockwise?

What determines the initial direction of motion?

Does the direction of rotation affect the period of the motion? Why or why not?

Click Reset. Change the mass to a larger value and leave the initial velocity set to –8 m/s.

Calculate what the value of period and frequency will be now.

period = ___________

frequency = ___________

Click Run and observe the motion for 3 seconds.

Does the mass of the ball have any effect on the period of motion? Why or why not?

Click Reset. Change the initial velocity to –16 m/s. Click Run to test your predictions.

How have the graphs changed (from the case with v = –8 m/s on page 151)? Why?

Click Reset. Set the initial velocity to 10 m/s and fill in the table on the next page by running with different rod lengths.

Period vs. Length of the Rod

length of rod	1.0 m	1.4 m	1.6 m	2.0 m
period				

Click Reset. Set the length of the rod to 1.6 m and fill in the following table by running with different initial velocities.

Period vs. Velocity of the Ball

velocity	6 m/s	8 m/s	10 m/s	12 m/s	16 m/s	20 m/s
period						

Plot the data from each table. You can use a graphing program if you have one. Sketch the results in the space below. *Do the shapes of the graphs make sense?*

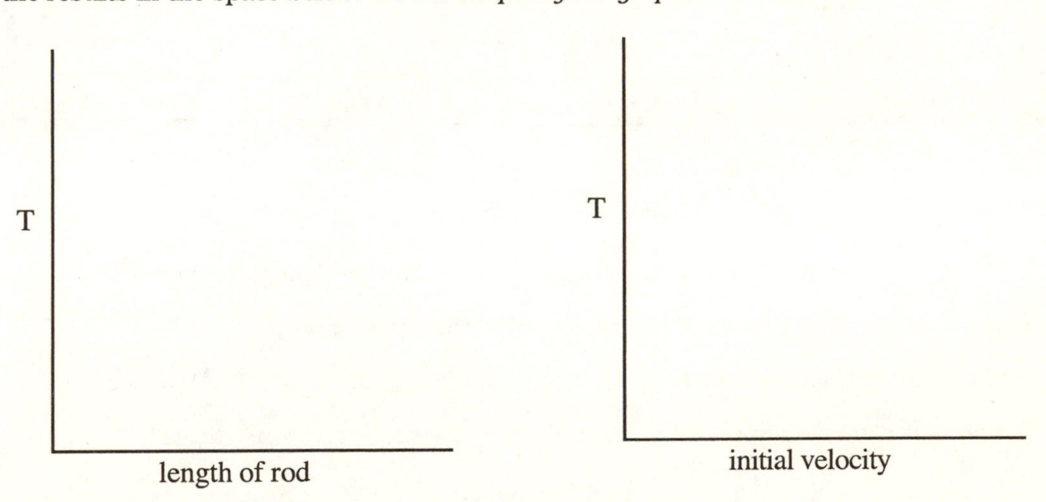

Self-Test Questions for Simulation 25

The questions below refer to these graphs showing the x component of the velocity of 4 different balls moving in circles with a constant speed. Two of the balls have a speed of 10 m/s. Each ball starts at the same initial position as in the simulation. The circles do not necessarily have the same radius.

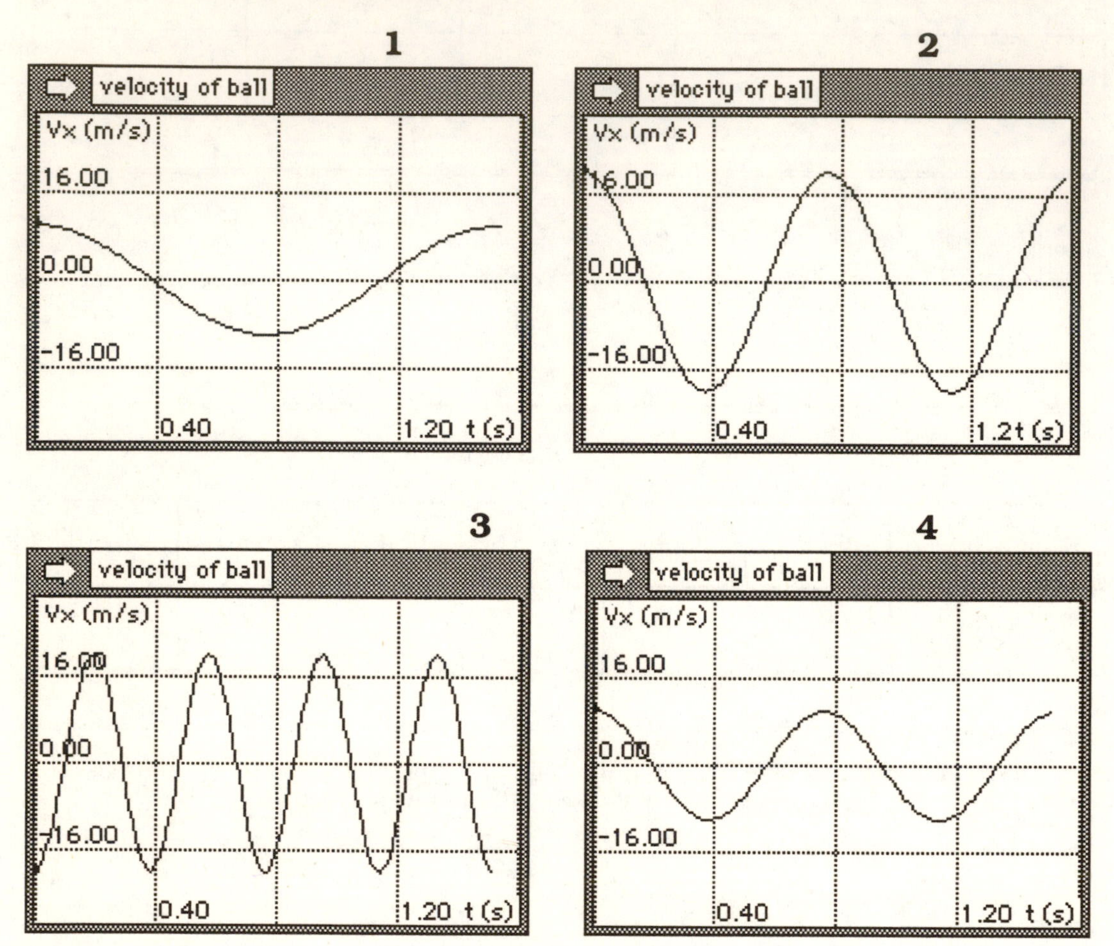

1. Which graphs represent balls with a speed of 20 m/s?
2. Which graph represents a ball moving in a circle with the highest frequency?
3. Which two graphs show the balls moving with the same period?
4. What is the approximate ratio of the radius of the motion of ball 2 to ball 4?
5. Which ball is moving in a circle in a direction opposite to the other three?

SIMULATION 26 A ROD ROTATING ABOUT ITS CENTER

Physics Review

When an object rotates about a fixed point, we can describe the motion with some simple equations. In particular for an object rotating with a constant angular acceleration (α), its angular velocity (ω) and angular displacement (θ) are given at any time t by the equations shown below. These equations are the rotational analogs of the linear equations relating acceleration (a), velocity (v), and position. Angular velocity is directly related to linear velocity as shown below.

Formulas you should be familiar with:

$$\alpha = \frac{\Delta\omega}{\Delta t} \qquad \omega = \omega_o + \alpha t \qquad \theta = \theta_o + \omega_o t + \tfrac{1}{2}\alpha t^2 \qquad v = r\omega$$

Reference topics: circular motion, angular velocity, angular acceleration

Simulation Details

When you run this simulation, the rod will rotate about a pivot point at its center. To get the rod up to a desired speed, it is accelerated from rest for 0.2 seconds. You can enter a value (between 0 and 40) for the angular acceleration in the box as indicated on your computer screen. Once the rod gets to the desired angular speed, it will rotate at a constant angular velocity for the remainder of the running time. You will have output meters for angular velocity (rad/s), linear velocity (m/s), and angular displacement (rad) for the two points labeled A and B on the rod. You can change the position of either point by using the sliders. The position of each point is measured relative to the center of the rod, which is about 6 meters long.

Calculate what value of angular acceleration is needed to get the rod rotating at 4 rad/s in 0.2 seconds.

α = ________

Set the angular acceleration to this value by typing it into the box. Click Run and observe the motion. The simulation will automatically pause at $t = 0.20$ seconds.

From the simulation data, record the following:

angular speed of the rod, ω = ____________

angular displacement of the rod, θ = _______

time = __________

Calculate what the angular displacement will be when the elapsed time is 1 second.

angular displacement, θ = _________

Hint

To use the equation $\theta = \theta_o + \omega t$, which holds for constant angular velocity, you must be careful to use the correct time (in this case $t = 1.0 - 0.2 = 0.8$) that it has that constant speed (ω) for, and you must use the correct initial angular displacement as well ($\theta_o = 0.40$ rad).

Click Run and the simulation will continue until $t = 1$ second when it will pause again.

From the simulation data, record the angular displacement and compare with your calculation above.

angular displacement, θ = ___________

How many revolutions (or part of) has the rod made at this point (t =1)?

Use the value of the angular speed that you recorded on the previous page to calculate the frequency of the rotation.

$$\text{frequency} = ______ \frac{\text{rad}}{\text{sec}} \times \frac{1 \text{ rev}}{2\pi \text{ rad}} = ________ \frac{\text{rev}}{\text{sec}}$$

which corresponds to a period of _________ seconds.

Click **Run** to continue running the simulation for 2 or 3 seconds. Use the tape player to find the time it takes for one complete cycle (at constant speed).

From the simulation data, record the period.

period = ________

How does this value compare with the one that you calculated above?

Click **Reset**. Set the position of point A to 3 m and the position of point B to 1 m.

Calculate the linear velocity of each point at t = 0.5 seconds.

linear velocity of point A = ________

linear velocity of point B = ________

Click **Run** and observe the motion for at least 0.5 seconds.

What is the ratio of distance of point A from the center to the distance of point B from the center?

:

What is the ratio of the angular velocity of point A to the angular velocity of point B at 0.5 seconds?

:

What is the ratio of the linear velocity of point A to the linear velocity of point B at 0.5 seconds?

:

What is the ratio of the angular velocity of point A to the angular velocity of point B at 0.1 seconds (i.e., the time when the rod is accelerating)?

:

What is the ratio of the linear velocity of point A to the linear velocity of point B at 0.1 seconds (i.e., the time when the rod is accelerating)?

:

Self-Test Questions for Simulation 26

The following questions refer to a rod like the one in the simulation.

1. True or false: When the angular acceleration of the rod is zero, its angular speed is constant.

2. True or false: The angular velocity of one end of the rod is the same as the angular velocity of the other end of the rod.

3. True or false: The linear velocity of a point on the rod depends on its distance from the center of the rod.

4. If the angular speed of a rotating rod increases from zero to 5 rad/s in 0.4 seconds, what is its angular acceleration?

5. What will be the angular displacement of that rod at t = 0.4 seconds?

SIMULATION 27 FORCES APPLIED TO A HINGED DOOR

Physics Review

Just as forces cause acceleration and hence changes in translational motion, torques cause changes in rotational motion. When you push on a door, you exert a torque on it and the door rotates about the axis through which it is hinged. To calculate the torque on a body about a point, you need to know the magnitude and direction of the force and the distance from the application point of the force to that point. To calculate the torque about point A for the force shown in CASE 1, you multiply the magnitude of the force by the distance L. For CASE 2, since the force is applied at an angle other than 90°, the sine of the angle comes into the equation and the torque is given by $FL\sin\phi$. There are two ways to arrive at this conclusion. In one way you break the force up into two components and see that only one ($F\sin\phi$) acting at a distance of L gives a nonzero contribution to the torque. The other way to look at this is to consider that the entire force F is acting at a distance $L\sin\phi$ from point A. In either case the torque is $FL\sin\phi$.

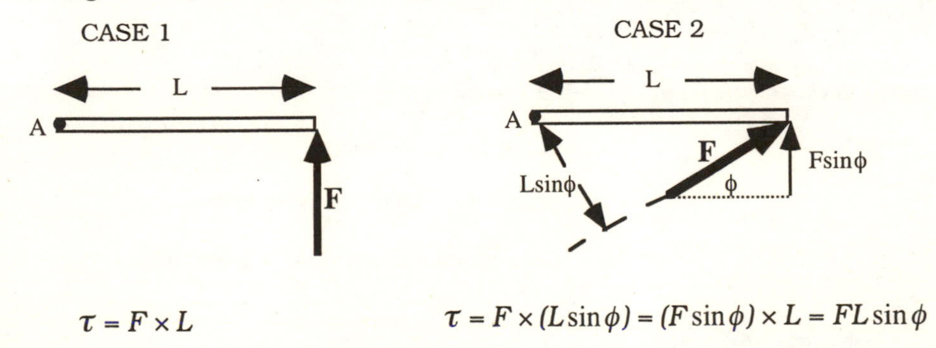

$\tau = F \times L$

$\tau = F \times (L\sin\phi) = (F\sin\phi) \times L = FL\sin\phi$

When more than one force is involved, the torque is calculated for each separately, and then they are added together, taking care to include the proper sign. Positive torques will result in counterclockwise rotation and negative torques produce clockwise rotation. This is the convention used in *Interactive Physics*.

Formulas you should be familiar with:

$\tau = FL\sin\phi$ $\quad\alpha = \dfrac{\Delta\omega}{\Delta t}$ $\quad\omega = \omega_0 + \alpha t$ $\quad\theta = \theta_0 + \omega_0 t + \frac{1}{2}\alpha t^2$

Reference topics: torque, angular acceleration

Simulation Details

In this simulation you will be looking down (from above) on a "revolving door." The door has a total width (you will see it as length) of 2.2 meters. There are two forces that can be applied to the door to cause it to rotate, and you can adjust their magnitude and direction (but not the point of application). Initially, point A is 1 m to the right of the center and point B is 0.5 m to the left of the center. The direction of the applied force is measured with respect to the x direction and therefore an angle of 90° results in a force that is applied perpendicular to the door, in the positive y direction, and an angle of 270° results in a force that is in the negative y direction. You can turn either force "off" by setting its magnitude to zero. This simulation will automatically pause after 3 seconds have elapsed.

Set the magnitude of the force at point B to zero. Set the magnitude and direction of the force at point A to 4 N and 90°. Click **Run** and observe for 3 seconds.

From the simulation data, record or calculate the following:

angular acceleration of the door = ________

angular velocity (ω) at 3 seconds = ________

angular displacement (θ) at 3 seconds = ______

Sketch the graph of the angular velocity of the door.

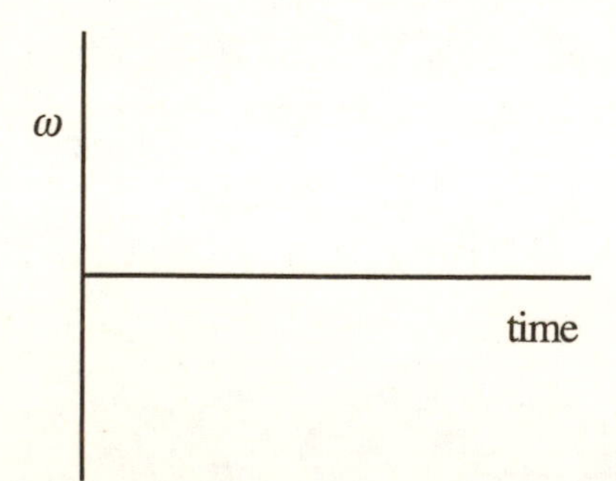

How many revolutions did the door make in 3 seconds?

How long did it take for the door to complete its first revolution?

How long did it take for the door to complete its second revolution?

How long did it take for the door to complete its third revolution?

Why is the time/revolution (period, sort of) decreasing as the door continues to rotate?

Assume that you were to decrease the magnitude of the force at point A to 2 N, leaving the direction fixed. How would the following quantities change? Circle the correct words.

- The torque would be (*1/2 as much, 1/4 as much, the same, doubled*).
- The angular acceleration would be (*1/2 as much, 1/4 as much, the same, doubled*).
- The angular velocity at 3 seconds would be (*1/2 as much, 1/4 as much, the same, doubled*).
- The angular displacement at 3 seconds would be (*1/2 as much, 1/4 as much, the same, doubled*).

Click **Reset**. Keep the magnitude of the force at point B set to zero. Change the magnitude of the force at A to 2 N, leaving the direction set at 90°. Click **Run** and observe the motion until the simulation pauses.

From the simulation data, record or calculate the following:

angular acceleration of the door = ________

angular velocity (ω) at 3 seconds = ________

angular displacement (θ) at 3 seconds = ________

Compare these values to those you recorded on the previous page. *Did things change as you expected?*

There is another way to achieve the exact same motion of the door (as the last case) by using a force exerted at point B instead of one at point A (i.e., F at A = 0.0). Calculate the magnitude needed for the force at B if its direction is set to 270°.

magnitude of force at point B = ___________

Set the magnitude of the force at A to zero and the direction of the force at B to 270°. Set the magnitude of the force at B to the value that you just calculated. Click **Run**.

From the simulation data, record or calculate the following:

angular acceleration of the door = ________

angular velocity (ω) at 3 seconds = ________

angular displacement (θ) at 3 seconds = _______

Compare these values to those you recorded on the previous page. *Is the motion the same?*

There is yet another way to get this motion of the door. It is by setting the magnitude of the force at point A to zero and the magnitude of the force at point B to 8 N. The force at B must then be applied at some angle. Calculate the value of this angle. Hint: there are two possible answers.

θ = __________

Click **Reset**. Set the magnitude of the force at point A to zero. Set the magnitude of the force at point B to 8 N and its angle to one of the values that you just calculated. Click **Run**.

From the simulation data, record or calculate the following:

angular acceleration of the door = ________

angular velocity (ω) at 3 seconds = ________

angular displacement (θ) at 3 seconds = ______

Is the motion exactly the same as in the prior case?

As you have seen, there are several ways to achieve the same motion of the door. Now we will explore ways to get the door into rotational equilibrium. Consider the situations given in the table, but before you run them, decide whether the door will be in equilibrium or not. REMEMBER TO RESET EACH TIME BEFORE CHANGING VALUES.

Force B		Force A		Equilibrium?
8 N	90°	4 N	90°	
4 N	90°	8 N	90°	
2 N	270°	4 N	90°	
4 N	180°	0 N	NA	
8 N	30°	2 N	90°	

Click **Reset**. Set the magnitude and direction of the force at point B to 5 N and 90° and the magnitude and direction of the force at point A to 3 N and 90°. Click **Run** and observe for about 1 second and then stop. DO NOT RESET. Change the magnitude of the force at point B to 6 N and click **Run**.

Explanation

Note that even though the door is moving, it is in rotational equilibrium in this case as well. The door has a constant angular velocity, zero angular acceleration and the net torque on it is zero. Objects in rotational equilibrium may or may not be moving. You should now see that there are many ways to get the door in rotational equilibrium. Theoretically there are an infinite number of combinations of the forces at A and B that will result in a net torque of zero and hence rotational equilibrium.

Self-Test Questions for Simulation 27

The questions apply to the two graphs showing the angular displacement and angular velocity of a door. In each graph there are two curves. The magnitude of the force for the curve numbered 1 is F and it is applied at an angle of 90° (*i.e.*, perpendicular to the door) at a distance L from its center.

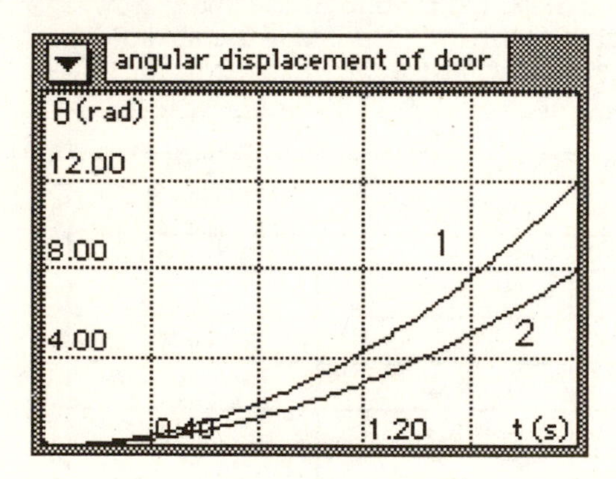

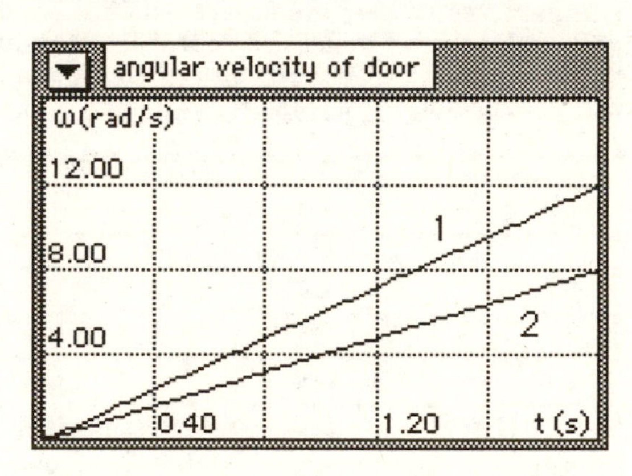

1. What is the approximate angular acceleration of the door represented by curve 1?
2. Approximately how many revolutions does the door represented by curve 2 make in the last second? (Remember there are about 6.28 radians in a revolution.)
3. If you keep the magnitude and direction of the force fixed, where must you apply it to achieve the data represented by curve 2?
4. If you keep the magnitude and point of application of the force fixed, at what angle must the force be applied to achieve the data represented by curve 2?
5. If you keep the direction of the force and its point of application fixed, what magnitude must the force have to achieve the data represented by curve 2?

SIMULATION 28 TORQUES ON A METER STICK

Physics Review

When an object is in rotational equilibrium, its angular acceleration is zero. As you should have seen at the end of the last simulation, the object is not necessarily at rest: it can be rotating at a constant angular speed. In this case we will be looking at a meter stick resting on a pivot with weights hanging from it, and we will be concerned with achieving a balanced state. If a mass hangs from the meter stick at a distance L from the center, then the torque about the center due to the hanging mass is just mgL, the product of its weight and the lever arm. Remember that to achieve complete equilibrium, the sum of the forces as well as the sum of the torques must be zero.

Reference topics: torques, equilibrium

Simulation Details

Three blocks hang from the meter stick. The meter stick is supported at its center and has a mass of 200 grams. A 50-gram block is suspended by a string from the left end of the meter stick and a 20-gram block is suspended 10 cm from the right end. The third block, which is located 20 cm to the right of the center of the meter stick, has an adjustable mass. The point about which all the torques are calculated is labeled as point A, and it is located initially at the center of the meter stick. You can adjust the location of this point by using the slider (–0.5 is at the left end and +0.5 is at the right end). According to *Interactive Physics* convention, torques that cause clockwise rotation are negative.

Calculate the torque caused by the 50-gram block and that of the 20-gram block about point A. Be sure to include the sign and record 4 digits after the decimal point.

torque of 50-gram block = __________

torque of 20-gram block = __________

In order to achieve rotational equilibrium, what must the torque of the third block be?

torque = ____________

What must the mass of the third block be to balance the meter stick?

mass = ______________

Set the mass of the third block to the value that you just calculated. Click **Run** and observe the motion.

From the simulation data, record or calculate the following:

torque due to 50-gram block = ____________

torque due to 20-gram block = ____________

torque due to third block = ____________

Compare these values to those that you just calculated. *Are they the same?*

Now suppose that you want to calculate the torques about a different point on the meter stick. Assume that this point A is now located 10 centimeters to the right of the center of the meter stick. In what follows, circle the correct word.

- The torque due to the 50-gram block will have a (*larger, smaller*) magnitude and be (*negative, positive*).
- The torque due to the 20-gram block will have a (*larger, smaller*) magnitude and be (*negative, positive*).
- The torque due to the third block will have a (*larger, smaller*) magnitude and be (*negative, positive*).
- The sum of these three torques will be (*positive, negative, zero*).

Click **Reset**. Leave the mass of the third block set to 85 grams. Set the location of point A to 0.1 meters using the slider. Click **Run**.

From the simulation data, record or calculate the following:

torque due to 50-gram block = ____________

torque due to 20-gram block = ____________

torque due to 85-gram block = ____________

sum of these three torques = ________________

Compare these values to your predictions on the previous page. *Do they agree?*

Is the meter stick in equilibrium?

Explanation

The meter stick is in equilibrium, which means that the net torque on it is zero. However, the sum of the torques caused by the three blocks is not zero. There are two other forces that contribute to the net torque on the meter stick that must be included. One is the force exerted on the meter stick by the support, and the other is the weight of the meter stick itself. The reason why these were not a factor on page 166 is because torques were calculated about the center and these forces act through the center of the meter stick resulting (in that case) in a zero contribution to the torque.

Calculate the value of the force exerted by the support on the meter stick.

force = _______

Calculate the torque of this force about point A (located 0.10 m to the left of the center).

torque = ___________

When this torque and the one due to the weight of the meter stick are included, the net torque should be zero.

torque due to 50-gram block = ___________

torque due to 20-gram block = ___________

torque due to 85-gram block = ___________

torque due to support force = ___________

torque due to weight of meter stick = ___________

NET TORQUE = ___________

Where must point A be located so that the torque due to the 50-gram block is zero?

Without changing the values of the masses of the blocks, can you find a location for point A so that the net torque is not zero?

Self-Test Questions for Simulation 28

These statements refer to the meter stick and blocks used in this simulation. True or false?

1. The sum of the forces can be zero even if the sum of the torques is not zero.
2. The torque exerted by any block about the center of the meter stick will be the same as the torque exerted by the same block about the end of the meter stick.
3. The force exerted on the meter stick by the support is not zero.
4. The net torque about the right end is the same as the net torque about the left end.
5. Using our sign convention, a block hung to the left of the center will result in a negative torque about the center.

SIMULATION 29 A MERRY-GO-ROUND (VECTORS)

Physics Review

Many objects move in circular or nearly circular paths. Planets around the sun, the Moon around the Earth, a point on a bicycle wheel, and a cart on a Ferris wheel are just a few of the many examples of circular motion. Forces and torques (τ) are responsible for keeping objects moving in a circle whether or not the speed is constant. Traditionally we speak of two components of the acceleration, one pointing inward (the centripetal acceleration, designated here by the symbol a_c) and one pointing along (or opposite to) the direction of motion (the tangential acceleration, designated here by the symbol a_t). The tangential acceleration is related to the angular acceleration (α) as given in the formulas below. The moment of inertia (I) is a constant of the problem determined by the mass and radius of the merry-go-round.

Formulas that you should be familiar with:

$$a_c = \frac{v^2}{r} \qquad \tau = I\alpha \qquad \alpha = \frac{a_t}{r}$$

Reference topics: circular motion, torque, angular acceleration, centripetal acceleration.

Simulation Details

In this simulation you will be able to adjust the torque on a merry-go-round. By *Interactive Physics* convention, positive torques will produce counterclockwise rotation and negative torques will produce clockwise rotation. As the merry-go-round rotates, you will observe various quantities associated with a teacup on its rim. You will be able to explore the relationship between the torque on the merry-go-round, the angular acceleration of the merry-go-round, the tangential acceleration of the teacup, and the centripetal acceleration of the teacup. It is not possible to display the two components of the acceleration in the traditional way (tangential and centripetal) in *Interactive Physics*, but you can figure out what is happening to each of them individually by looking at the numerical values shown on the meters.

Set the torque to the maximum negative value. Click **Run** and observe the simulation for about 5 seconds.

In what direction is the merry-go-round rotating? Clockwise or counterclockwise?

Click **Reset**. Set the torque to a positive value, but not the maximum positive value. Click **Run** and observe the motion for about 10 seconds. Use the tape player to answer these questions.

In what direction is the merry-go-round rotating? Clockwise or counterclockwise?

What is happening to the acceleration vector of the teacup as the simulation progresses? Is it increasing, decreasing, or remaining the same size? Is its direction changing? What is the direction approaching?

What is happening to the value of the angular acceleration of the merry-go-round as the simulation progresses? Is it increasing, decreasing, or remaining the same size?

What is happening to the tangential acceleration of the teacup as the simulation progresses? Is it increasing, decreasing, or remaining the same size? Why?

From the simulation data, record or calculate the following:

tangential acceleration = _______

Click **Reset**. Change the value of the torque to the maximum positive value. Do not run yet.

Predict whether the magnitude of the tangential acceleration of the teacup will be larger, smaller, or remain the same.

Click **Run** to test your prediction.

From the simulation data, record or calculate the following:

tangential acceleration = __________

Did you predict that it would be larger?

Click **Reset**. Set the torque to some positive value. Click **Run** and observe for about 10 seconds.

What is happening to the centripetal acceleration of the teacup? Is its magnitude changing? Can you explain why?

Can you figure out a way to keep the centripetal acceleration of the teacup constant?

Hint

Set the torque to some positive value and run until you reach a point where the centripetal acceleration is at least 2 m/s^2. Then stop the simulation. Do not reset. Adjust what you need to (using sliders) to make the centripetal acceleration constant from that point on. Then run to see if it is constant.

What is the direction of the acceleration vector now? Can the acceleration vector have that direction if the torque is not zero?

What are the units of angular acceleration?

What are the units of tangential acceleration?

What are the units of centripetal acceleration?

Self-Test Questions for Simulation 29

Suppose a torque is being applied to a uniform wheel of some kind. Which of the following statements are true and which are false?

1. A constant torque will produce a constant angular acceleration.
2. A positive torque will produce an increasing angular acceleration.
3. A zero torque will produce a zero angular acceleration.
4. A negative torque, applied to a wheel initially at rest, will produce a centripetal acceleration that is increasing.
5. A constant centripetal acceleration is caused by a constant torque.

SIMULATION 30 A MERRY-GO-ROUND (GRAPHS)

Physics Review

This simulation is similar to the previous one. Remember the relevant quantities are torque on the merry-go-round (τ), angular acceleration (α) and angular velocity (ω) of the merry-go-round, tangential acceleration (a_t), centripetal acceleration (a_c), and tangential velocity (v) of the teacup. The centripetal acceleration is a function of tangential velocity and always has a positive magnitude and a direction of inward, toward the center of the circle. Since the velocity is squared here, it does not matter in what direction the merry-go-round is rotating, the centripetal acceleration will have the same value if the teacup is moving at 1 m/s clockwise or counterclockwise. The angular acceleration is directly related to the torque, so a constant torque means a constant angular acceleration, which for the teacup means a constant tangential acceleration as well. As in the case of linear motion, the tangential velocity and tangential acceleration are related as shown below. The moment of inertia (I) is a constant of the problem determined by the mass and radius of the merry-go-round.

Formulas you should be familiar with:

$$a_c = \frac{v^2}{r} \qquad \tau = I\alpha \qquad \alpha = \frac{a_t}{r} \qquad v = v_o + a_t t \qquad \omega = \omega_o + \alpha t$$

Reference topics: circular motion, torque, angular acceleration, centripetal acceleration.

Simulation Details

In this simulation the torque on the merry-go-round is adjustable. You will predict and study the graphs of angular velocity of the merry-go-round, tangential acceleration, centripetal acceleration and tangential velocity of the teacup. You can change the value of the torque in real time (i.e., when the simulation is still running).

Set the value of the torque to a positive value.

Predict what the graphs of angular velocity, tangential acceleration, centripetal acceleration, and tangential velocity will look like. Sketch them in the area below.

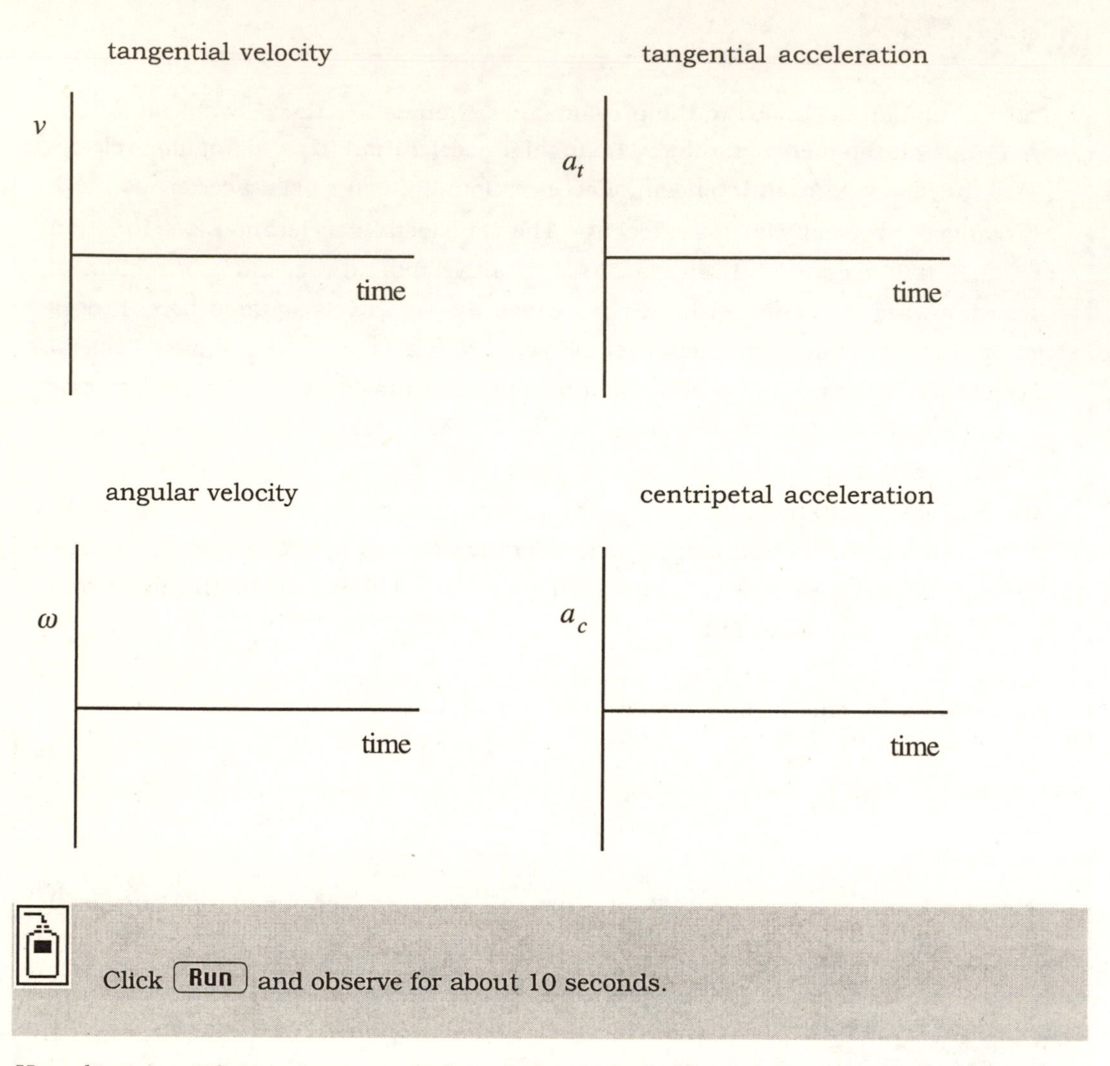

Click **Run** and observe for about 10 seconds.

How do your predictions compare to the graphs shown in the simulation?

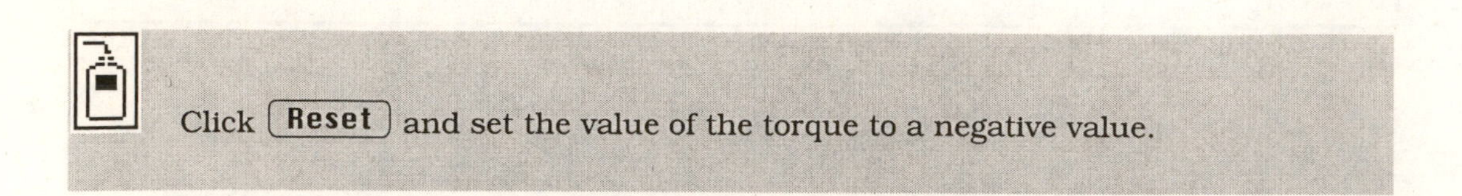

Click **Reset** and set the value of the torque to a negative value.

Predict what the graphs of angular velocity, tangential acceleration, centripetal acceleration, and tangential velocity will look like. Sketch them in the area below.

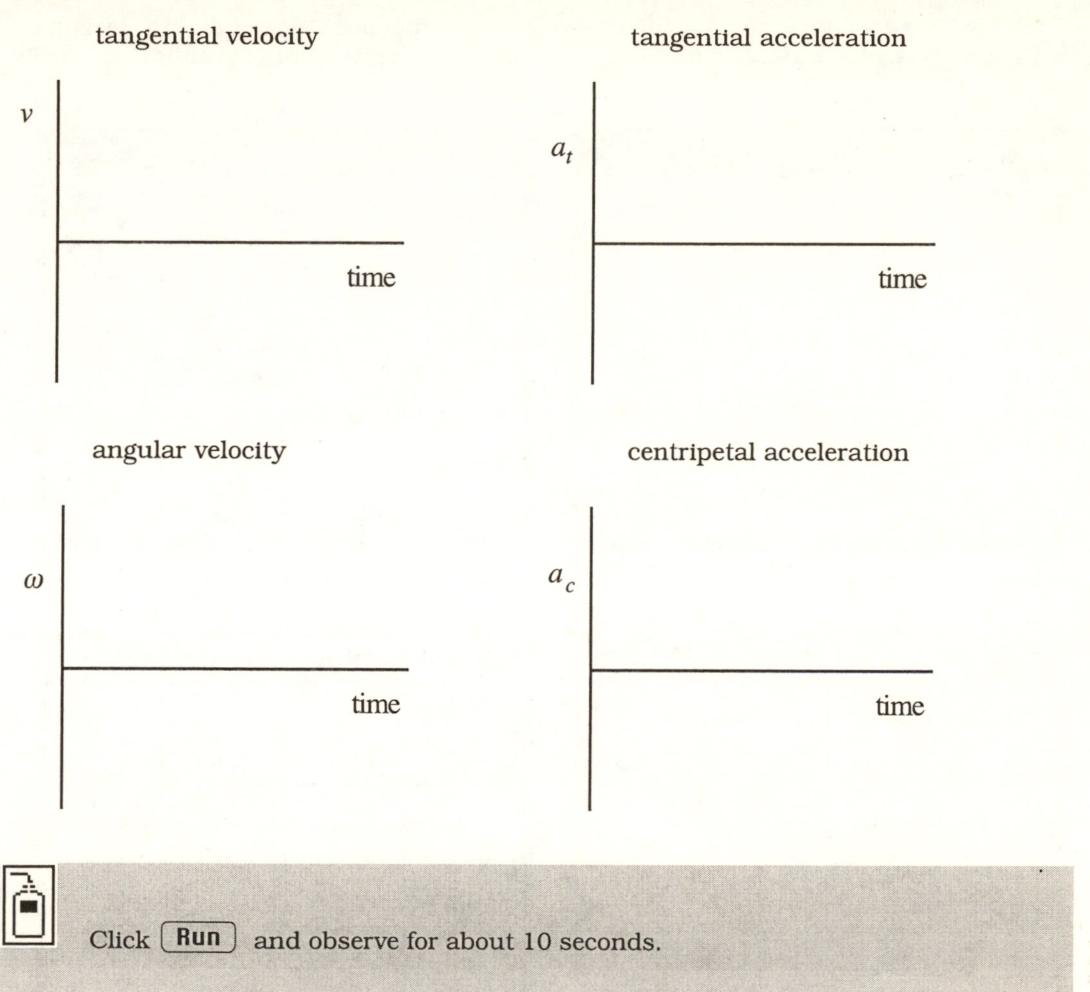

Click **Run** and observe for about 10 seconds.

How do your predictions compare to the graphs shown in the simulation?

Click **Reset** and set the torque to a positive value and run for about 5 seconds. Stop and then set the torque to zero and run for 3 seconds. What must you now do to get the merry-go-round to slow down and reverse direction? Try it.

Self-Test Questions for Simulation 30

Each of the graphs represents a different variable for the merry-go-round or a teacup on the edge of the merry-go-round. All the questions will apply to these graphs.

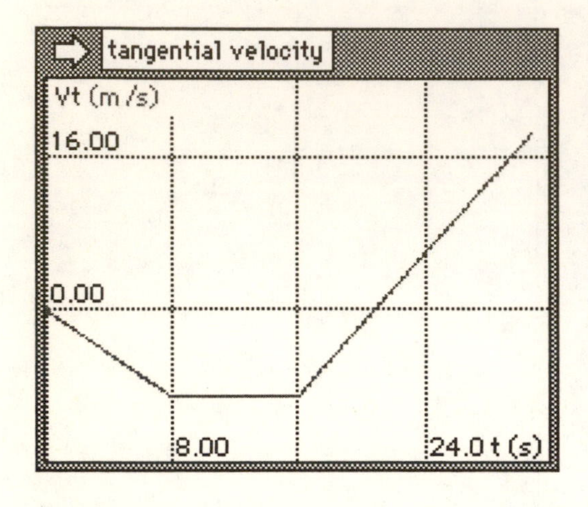

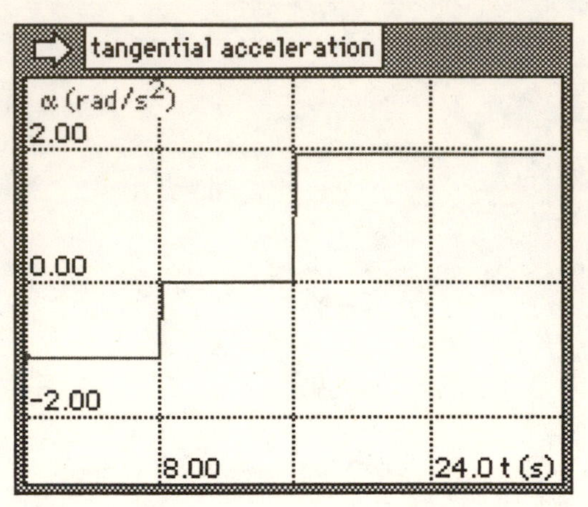

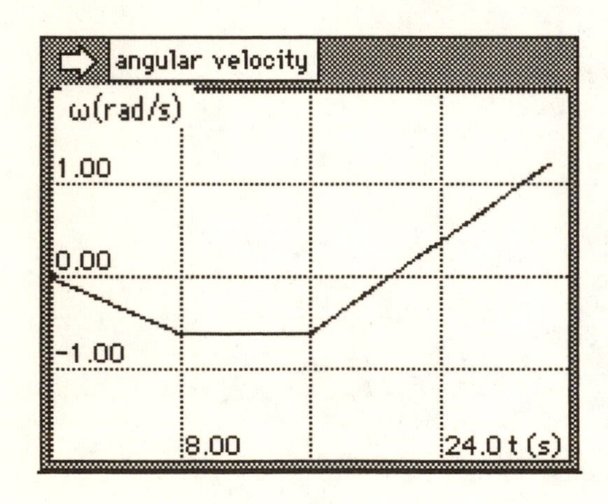

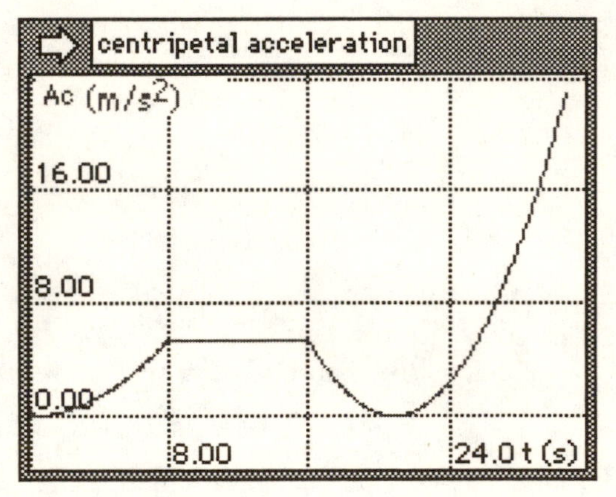

1. Is the torque positive or negative during the first 8 seconds?

2. How many times does the merry-go-round reverse direction?

3. At what time(s) does the merry-go-round reverse direction?

4. Is the speed of the teacup increasing or decreasing during the first 8 seconds?

5. During what time period is the speed of the teacup increasing at the fastest rate?

SIMULATION 31 A BLOCK OSCILLATING ON A SPRING (VECTORS, EQUATIONS)

Physics Review

A block attached to a spring, when displaced from the unstretched position (known as equilibrium position) will execute oscillatory motion about that position. In the simplest model the force that the spring exerts on the block is given at any instant by the product of the spring constant (k) and the amount of stretch (x) and is opposite to the direction of the displacement. This is known as Hooke's law. At any instant in time the block has an acceleration, a velocity, and a position given by sine or cosine functions (depending on the initial conditions) of the form shown below. The values of position, velocity, and acceleration of the block depend on the period of the motion (T), which in turn depends on the spring constant and the mass (m) of the block.

Formulas you should be familiar with:

$$F = -kx \qquad x = A\cos\frac{2\pi t}{T} \qquad T = 2\pi\sqrt{\frac{m}{k}}$$

Reference topics: simple harmonic motion, period, amplitude, Hooke's law

Simulation Details

In this simulation there is a block with an adjustable mass attached to a spring with an adjustable spring constant. The block will slide back and forth in a horizontal direction as the spring gets compressed and extended. The unstretched length of the spring is 3 meters. You can set the initial conditions to a compressed spring (negative values of compression/extension) or an extended spring (positive values of compression/extension). Position, velocity, and acceleration graphs will allow you to study the motion and determine the period and equations of motion. There is no friction in this simulation. This simulation will pause when 3 seconds have elapsed.

Set the mass to 0.5 kg, the compression/extension to 0.5 m and the spring constant to 10 N/m. Click Run.

From the simulation data, record or calculate the following:

period of the motion = ____________

amplitude of the motion = _________

maximum acceleration of the block = _________

Now calculate the period, amplitude, and maximum acceleration from the values of mass, spring constant, and initial compression/extension.

period of the motion = ____________

amplitude of the motion = _________

maximum acceleration of the block = _________

How do your calculations compare with the values that you recorded from the simulation?

Now you will study the acceleration and velocity vectors. We will look at four time regions separately. Use the [O O] to find the regions of interest. In what follows, circle the appropriate answer or fill in the value from the simulation.

During the first quarter of a cycle ($t < T/4$):

- The block is moving to the (*left, right*).
- Its acceleration vector is getting (*larger, smaller*).
- Its velocity vector is getting (*larger, smaller*).
- Draw the velocity and acceleration vectors in the middle of this quarter cycle.

During the second quarter of a cycle ($T/4 < t < T/2$):

- The block is moving to the (*left, right*).
- Its acceleration vector is getting (*larger, smaller*).
- Its velocity vector is getting (*larger, smaller*).
- Draw the velocity and acceleration vectors in the middle of this quarter cycle.

During the third quarter of a cycle ($T/2 < t < 3T/4$):

- The block is moving to the (*left, right*).
- Its acceleration vector is getting (*larger, smaller*).
- Its velocity vector is getting (*larger, smaller*).
- Draw the velocity and acceleration vectors in the middle of this quarter cycle.

During the last quarter of a cycle ($3T/4 < t < T$):

- The block is moving to the (*left, right*).
- Its acceleration vector is getting (*larger, smaller*).
- Its velocity vector is getting (*larger, smaller*).
- Draw the velocity and acceleration vectors in the middle of this quarter cycle.

During what part(s) of the cycle (i.e., at what time—half cycle, quarter cycle, etc.) is the acceleration zero?

During what part(s) of the cycle (i.e., at what time—half cycle, quarter cycle, etc.) is the velocity zero?

Click **Reset**. Leave the compression/extension set to 0.5 m and change the spring constant to 22 N/m and the mass to 5.5 kg. Click **Run** and observe the motion for about 8 seconds.

Now you will determine the equations for position, velocity, and acceleration. In each case you will be given four choices and you must circle the one that is correct. Use the tape player to help you as you study the graphs in the simulation.

Mathematics Help

These angles are all in radians.

$\cos(0) = \cos(2\pi) = 1$ $\cos(\pi) = -1$ $\cos\left(\frac{\pi}{2}\right) = 0$

$\sin(0) = \sin(\pi) = \sin(2\pi) = 0$ $\sin\left(\frac{\pi}{2}\right) = 0$

$$x = A\cos\frac{2\pi t}{T}$$

When you plug in a value for t in this equation, the angle is in radians. If you have a calculator that can do angles in radians (and therefore take the sine or cosine directly), fine. But if your calculator cannot handle radians, then you must convert this number of radians into degrees and then take the cosine.

Conversion: 1 rad = 57.3°.

Although knowing how to take a derivative will make the following a bit easier, it is entirely possible to do the following without knowing calculus. Hint: Look at the curves.

Circle the proper equation for position.

$x = 0.5\cos(2t)$	$x = -0.5\sin(2t)$
$x = -0.5\cos(2t)$	$x = 0.5\sin(\pi t)$

Circle the proper equation for velocity.

$v = \sin(\pi t)$	$v = -\sin(2t)$
$v = \sin(2t)$	$v = \cos(2t)$

Circle the proper equation for acceleration.

$a = 2\cos(\pi t)$	$a = 2\cos(2t)$
$a = -2\cos(2t)$	$a = 2\sin(2t)$

Use the simulation data for t = 2 seconds to check your circled answers.

Click **Reset**. Change the compression/extension to –0.5 m and leave the spring constant set at 22 N/m and the mass set at 5.5 kg. Click **Run** and observe the motion for about 8 seconds.

The period and amplitude of this motion is the same as the previous motion. However, the equations of motion will be different, because in this case the spring was initially compressed rather than initially extended. Look back at the equations for position, velocity, and acceleration in the previous section and find the ones that correspond to this motion, and write them in the space below.

$x =$

$v =$

$a =$

Use the simulation data for $t = 2$ seconds to check your answers.

Self-Test Questions for Simulation 31

The questions below will refer to these two graphs. Each curve in each graph represents one of two blocks (with different masses) oscillating in simple harmonic motion. In both cases the spring constant and the amplitude (A) are the same.

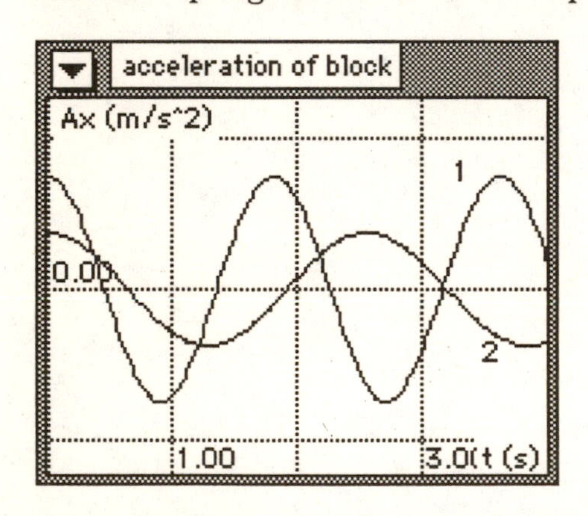

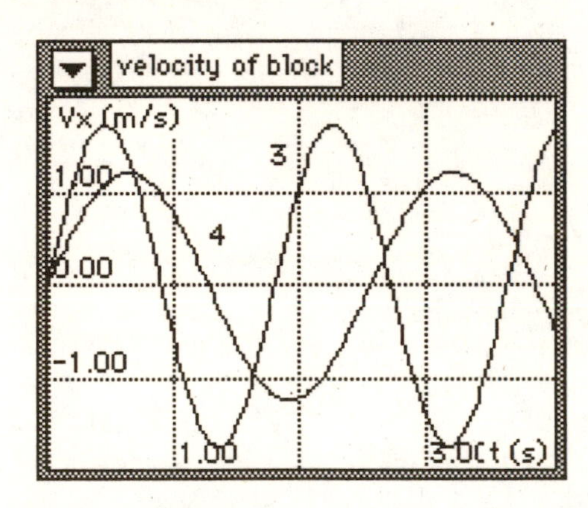

1. Which velocity curve goes with acceleration curve 1?
2. Which acceleration curve represents the block with the smaller mass?
3. What is the approximate ratio of the masses?
4. Which block has the greater speed at 3 seconds?
5. Which block experiences the greater maximum force?

SIMULATION 32 A BLOCK OSCILLATING ON A SPRING (ENERGY AND DAMPING)

Physics Review

Another way to gain insight into simple harmonic motion is by studying energy. The spring, when compressed or stretched a distance x from its equilibrium length, has potential energy. The initial compression (or extension) from equilibrium is equal to the amplitude of the motion (A), and the total available energy for the system (block and spring) is equal to $\frac{1}{2}kA^2$. This energy gets transferred from the spring to the block and back again (twice) in a cycle characterized by the period of the motion. So long as no dissipative forces are present, the total mechanical energy remains constant and we can study how the energy changes from kinetic to potential. If there are dissipative forces present, such as friction, the total mechanical energy decreases over time and the oscillation eventually stops. As long as the block is moving, the coefficient of kinetic friction is used to calculate the force of kinetic friction. When the block finally stops, the coefficient of static friction will determine the maximum value of the static frictional force.

Formulas you should be familiar with:

$$KE_{block} = \frac{1}{2}mv^2 \qquad PE_{spring} = \frac{1}{2}kx^2 \qquad T = 2\pi\sqrt{\frac{m}{k}} \qquad f = \frac{1}{T}$$

Reference topics: simple harmonic motion, energy, friction

Simulation Details

In this simulation as in the last, there is a block attached to a spring. The block will slide back and forth in a horizontal direction as the spring gets compressed and extended. In this case, however, you will study how the energy of the system varies over time. You will be able to see how things change if the block oscillates on a surface that is not smooth (i.e., friction is involved). You can set the coefficient of friction to any value between 0 and 1 with the slider, and the coefficient of static friction is automatically set to that value plus 0.1. The initial amplitude is determined by setting the initial compression/extension of the spring.

Positive values indicate springs that are initially extended from their equilibrium position. This simulation will pause when 3 seconds have elapsed.

Set the spring constant to 30 N/m, the mass to 3.0 kg, the coefficient of friction to 0.0 and the compression/extension to 1.0. Click Run.

From the simulation data, record or calculate the following:

period of the motion = ____________

frequency of the oscillation = ____________

amplitude of the motion = ____________

total mechanical energy of the system = ____________

maximum kinetic energy of the block = ____________

maximum potential energy of the spring = ____________

maximum speed of the block = ____________

In what follows, circle the appropriate answer or fill in the value from the simulation. Remember, to get numerical values just click on the arrow in the top left corner of the graph

During the first quarter of a cycle ($t < T/4$):

- The block is moving to the (*left, right*).
- The kinetic energy of the block starts out with a value of ____________ and (*increases, decreases*) to a value of ____________.
- The potential energy of the spring starts out with a value of ____________ and (*increases, decreases*) to a value of ____________.
- The total mechanical energy of the system (*remains constant, is changing*).

During the second quarter of a cycle ($T/4 < t < T/2$):

- The block is moving to the (*left, right*).
- The kinetic energy of the block starts out with a value of ____________ and (*increases, decreases*) to a value of ____________.
- The potential energy of the spring starts out with a value of ____________ and (*increases, decreases*) to a value of ____________.
- The total mechanical energy of the system (*remains constant, is changing*).

During the third quarter of a cycle ($T/2 < t < 3T/4$):

- The block is moving to the (*left, right*).
- The kinetic energy of the block starts out with a value of ____________ and (*increases, decreases*) to a value of ____________.
- The potential energy of the spring starts out with a value of __________ and (*increases, decreases*) to a value of ____________.
- The total mechanical energy of the system (*remains constant, is changing*).

During the last quarter of a cycle ($3T/4 < t < T$):

- The block is moving to the (*left, right*).
- The kinetic energy of the block starts out with a value of ____________ and (*increases, decreases*) to a value of ____________.
- The potential energy of the spring starts out with a value of __________ and (*increases, decreases*) to a value of ____________.
- The total mechanical energy of the system (*remains constant, is changing*).

Use the to move to the time where one cycle has been completed.

How many "cycles" has the kinetic energy graph been through?

How many "cycles" has the potential energy graph been through?

Click Reset. Change the mass to 2 kg and leave the spring constant set to 30 N/m and the coefficient of kinetic friction set to 0.0. Do not run yet.

Predict how any of the following will or will not be different. *Will they be the same, larger or smaller than the previous case with the mass set to 3 kg?* In what follows, circle one.

- The period of the motion will be (*larger, smaller, the same*).
- The frequency of the oscillation will be (*larger, smaller, the same*).
- The amplitude of the motion will be (*larger, smaller, the same*).
- The total mechanical energy of the system will be (*larger, smaller, the same*).

- The maximum kinetic energy of the block will be (*larger, smaller, the same*).
- The maximum potential energy of the spring will be (*larger, smaller, the same*).
- The maximum speed of the block will be (*larger, smaller, the same*).

Click **Run** to test your predictions.

From the simulation data, record or calculate:

period of the motion = ____________

frequency of the oscillation = ____________

amplitude of the motion = __________

total mechanical energy of the system = __________

maximum kinetic energy of the block = __________

maximum potential energy of the spring = ________

maximum speed of the block = ________

Compare these values to those that you recorded on page 184. *How do your predictions compare with how things actually changed?*

Click **Reset**. Change the spring constant to 20 N/m and leave the mass set to 2 kg and the coefficient of kinetic friction set to 0.0.

Predict how any of the following will or will not be different. *Will they be the same, larger or smaller than the previous case with the spring constant set to 30 N/m?*

- The period of the motion will be (*larger, smaller, the same*).
- The frequency of the oscillation will be (*larger, smaller, the same*).
- The amplitude of the motion will be (*larger, smaller, the same*).
- The total mechanical energy of the system will be (*larger, smaller, the same*).
- The maximum kinetic energy of the block will be (*larger, smaller, the same*).

- The maximum potential energy of the spring will be (*larger, smaller, the same*).
- The maximum speed of the block will be (*larger, smaller, the same*).

Click [Run] to test your predictions.

From the simulation data record or calculate:

period of the motion = ____________

frequency of the oscillation = ____________

amplitude of the motion = ________

total mechanical energy of the system = __________

maximum kinetic energy of the block = __________

maximum potential energy of the spring = ________

maximum speed of the block = ________

Compare these values to those that you recorded on page 186. *How do your predictions compare with how things actually changed?*

Click [Reset]. Leave the mass set to 2 kg and the spring constant set to 20 N/m. Change the coefficient of kinetic friction to 0.15. Click [Run] and continue running until the block stops moving (i.e., its speed is 0.0). Since the simulation automatically pauses at 3 seconds, you may have to click [Run] to go on from there if the speed of the block is not yet zero.

Sketch the graphs of potential, kinetic, and total energy that are shown in the simulation all on the same graph. Use colored pencils or different types of lines to indicate which graph is which. Mark each graph with an X at the point where the block comes closest to its initial position.

time

How close does it come to its original position?

What is happening to the total mechanical energy of the system? Why?

Explanation

What has changed from the case without friction? Just about everything! The period and frequency of the motion are no longer defined, as the motion does not repeat itself. The amplitude is also not defined as it is decreasing over time. The maximum kinetic energy that the block attains is now less than the total initial energy, and therefore the maximum speed of the block is also less than it was. Also, the point where the block has its maximum speed is no longer at $x = 0$. The only thing that has not changed is the maximum value of the potential energy, which is stored in the spring at $t = 0$.

Use the tape player to locate the place where the block eventually comes to rest.

At that point, record or calculate the following:
(the coefficient of static friction is set at 0.1 more than the coefficient of kinetic friction)

stretch or compression of the spring = ___________

total mechanical energy of the system = __________

potential energy of the spring = ___________

total force on the block = __________

force exerted by the spring on the block = _________

force of static friction on the block = __________

Explanation

Notice that the block does not stop at the equilibrium position. It is not simple to calculate where the block will stop. However, when it stops, the net force on the block is zero and its velocity is zero. The force exerted by the spring (kx) must be equal to the force of static friction. If you calculated the force of static friction in the last part, you may have found that it was larger than kx. Remember that for static friction (when the object is not moving), $\mu_s N$ gives the maximum value that the force of static friction can attain and in this case it has a value less than that maximum (equal and opposite to kx).

From energy considerations you should be able to check that the total mechanical energy lost by the block/spring system is equal to the work done by the force of friction.

initial energy − final energy = energy lost

_____________ − _________________ = _________

The force of friction on the block is = ____________

and the work done by the force of friction is equal to that force times the distance that the block travels. So the big question is, how far does the block travel? Use the [O O] to go back through the motion and find the total distance traveled by the block.

Then the work done by the force of friction = ____________ × __________ = ___________

This should be equal to the energy lost by the block/spring system that you calculated above.

Self-Test Questions for Simulation 32

The questions below will refer to these two graphs. They each show the kinetic energy, potential energy and the total energy of a block attached to a spring sliding on a horizontal surface with friction. One block is more massive than the other, and they are released from rest with the same initial amplitude.

Block A

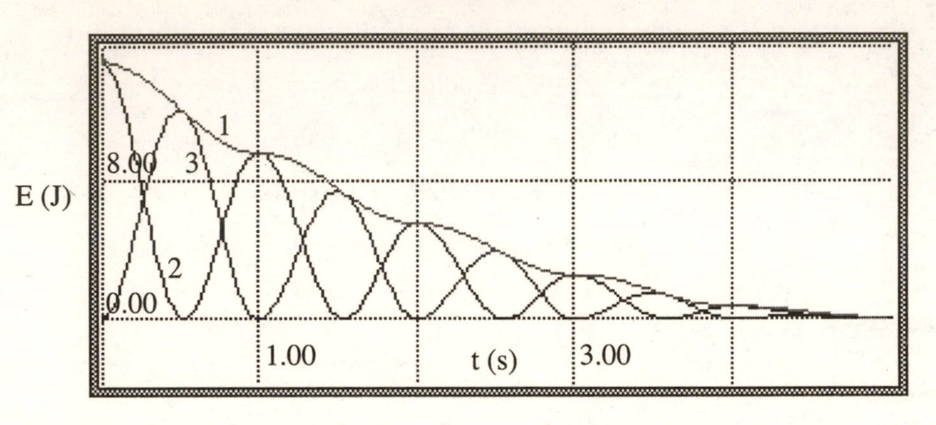

Block B

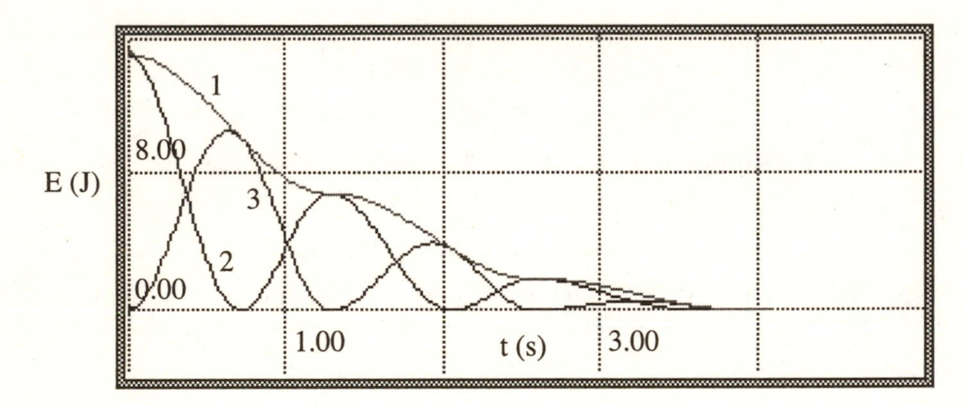

1. About how long does it take for the block in graph B to stop?
2. Which spring has a greater spring constant, or are they the same?
3. Which numbered curve represents the kinetic energy of a block?
4. Which graph represents the block with a greater mass?
5. Is there ever a time where the kinetic energy of a block is equal to the potential energy of its spring?

SIMULATION 33 SIMPLE PENDULUM

Physics Review

Another type of periodic motion is that of a simple pendulum, a mass hanging at the end of a rope of length L. The pendulum bob is released from some initial angular displacement (θ_o), swinging down through its equilibrium position (had it not been displaced) and then on to the other side and back again. In the absence of any air resistance (or other dissipative forces) it will continue to repeat this exact motion over and over again. For small initial angular displacements, the period of the motion (T) is given by the formula shown below and the motion is simple harmonic. The equation of motion is analogous to that of a mass on a spring (see simulation 31), except the displacement (θ) replaces the displacement x and ω depends on different parameters. You can analyze the motion with the help of the figure shown below (F is the tension in the rope). Applying Newton's laws perpendicular to and in the direction of motion gives the two equations below, which will hold at any angle θ. The two gray spheres represent the bob at the maximum swing and the lowest point.

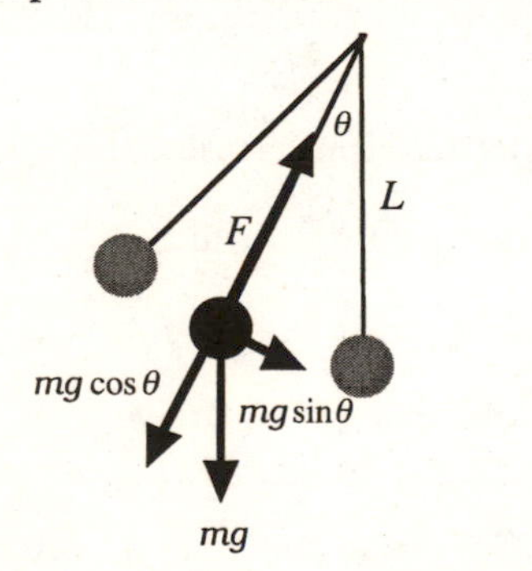

$$F - mg\cos\theta = ma_c = m\frac{v^2}{L}$$

$$mg\sin\theta = ma_t$$

Formulas you should be familiar with:

$$T = 2\pi\sqrt{\frac{L}{g}} \qquad a_c = \frac{v^2}{r} \qquad KE = \tfrac{1}{2}mv^2 \qquad PE = mgh$$

$$\theta = \theta_o \cos\omega t \qquad \omega = \sqrt{\frac{g}{L}}$$

Reference topics: simple harmonic motion, simple pendulum

Simulation Details

The mass of the bob is adjustable as is the length of the rope. The initial angular displacement is adjustable and is measured in degrees from the equilibrium position of the bob (i.e., the negative y direction). Positive angles will result in an initial displacement to the right. Output meters give simulation data for the tension in the rope as well as the energy (kinetic, potential and total mechanical) of the bob. The velocity and acceleration vectors are shown as well as graphs of velocity (positive values for the bob moving to the right and negative values for the bob moving to the left) and the magnitude of the acceleration (always positive).

Set the mass of the bob to 10 kg, the length of the rope to 4 m and the initial displacement to 10°. Click **Run** and observe the motion until the bob reaches its lowest point and then stop.

Use the [tape] to find the time when it is as close to its low point as you can get. From the simulation data, record or calculate the following:

the change in the potential energy (this point – initial) = ________

the change in the kinetic energy (this point – initial) = __________

the velocity as it passes through this low point = ________

the tension in the rope = _______

Use the energy data to calculate the velocity at that point. Then use the velocity data to calculate the tension in the rope. You should get values close to those above.

velocity = ___________

tension = __________

Calculate the value of the acceleration at the low point. It is just the centripetal acceleration $a_c = \frac{v^2}{r}$.

calculated acceleration = ___________

acceleration shown in simulation = _______________

Calculate what the tension, acceleration, and velocity will be when the pendulum swings to the other side (i.e., angular position of -10°).

tension = ___________

acceleration = _______

velocity = ___________

Click **Run** to see the pendulum swing to the right side. Stop close to that point.

From the simulation data, record or calculate the following:

the tension at maximum swing = __________

the acceleration at maximum swing = ____________

the velocity at maximum swing = _____________

Compare the values for tension, acceleration, and velocity to those that you just calculated. They should be the same (or very close). If they are not, read the explanation on the next page and then repeat the calculation.

Explanation

At this point of maximum swing, the velocity of the bob is zero. The velocity reached a maximum value when the bob went through the low point, and it has been decreasing ever since. Since the velocity is zero, the centripetal acceleration at this point is zero, so the net force in the radial direction is also zero. Hence the tension (F) is equal to $mg\cos\theta$. Note that this value is smaller than the value of the tension at the low point (which is the maximum value). The acceleration at this point is purely tangential (i.e., in the direction of motion) and equal to $g\sin\theta$.

Click [Run] to see the pendulum swing back to the low point. Stop as close to that point as possible.

From the simulation data, record or calculate the following:

the acceleration of the bob = __________

How does this compare to the value that you calculated before, when the bob went through going the other way (i.e., on the previous page)?

What is the maximum value of the acceleration of the pendulum and where does it occur?

What is the minimum value of the acceleration and where does it occur?

Explanation

Unlike the velocity, the acceleration of the pendulum is never zero and its direction is complicated as well. For small initial displacements, it has a maximum value at the end points. Here it is directed perpendicular to the rope (i.e., completely tangential). At the low point the acceleration is a minimum and is directed parallel to the rope (i.e., completely centripetal). At all other points the acceleration is the vector sum of the two components. See how the vector changes in both magnitude and direction!

Click Run to let the pendulum complete at least one full swing.

From the simulation data, record or calculate the following:

the period = ___________

Calculate the value of the period and compare to the simulation result.

period = ___________

If this is simple harmonic motion, then the equation for angular displacement is $\theta = \theta_0 \cos \omega t$ and the equation for velocity is $v = -v_0 \sin \omega t$. Pick <u>any time</u> and calculate displacement and velocity using these equations. Enter your results in the table along with values shown in the simulation. The maximum velocity is v_0 (m/s) and the maximum displacement is θ_0 (radians).

Mathematics Help

Remember, ω is in rad/sec, so that when you multiply it by the time, you will have an angle in radians. If you do not have a calculator that can accept angles in radians, then you must convert the number of radians into degrees and then take the sine. Conversion: 1 rad = 57.3°.

time =	velocity	displacement
simulation data		
calculated from equation		

Sketch the velocity and acceleration curves for one complete cycle in the space below. You will need to refer back to them for comparison, so include some information about time and maximum and minimum values.

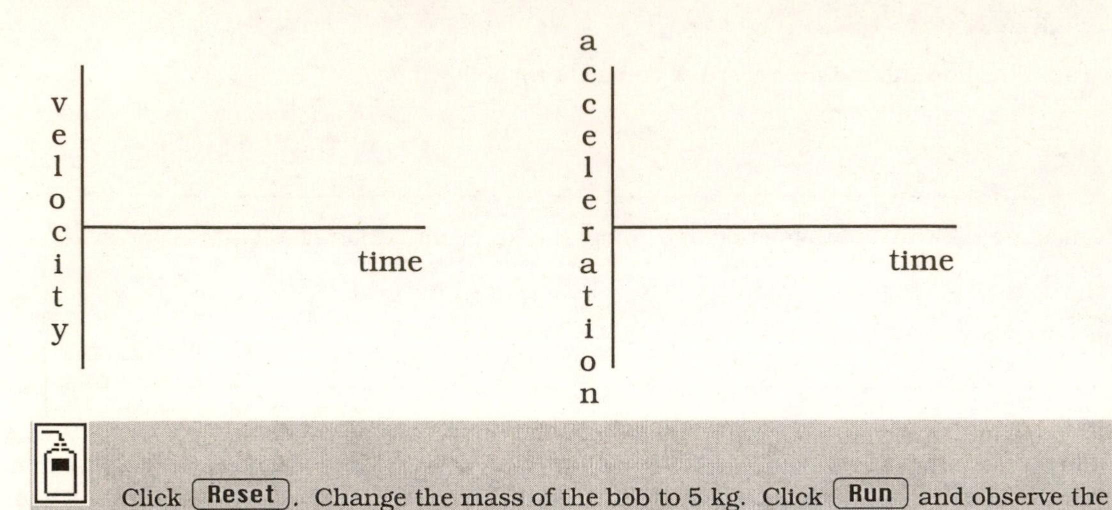

Click Reset. Change the mass of the bob to 5 kg. Click Run and observe the motion for one cycle.

Compare the velocity and acceleration curves to the ones you just sketched for a mass of 10 kg. *What effect did changing the mass have on the velocity and acceleration curves?*

What effect did changing the mass have on the period?

What effect did changing the mass have on the tension and the energies?

From the simulation data, record the maximum tension, the minimum tension, the period, the maximum acceleration, the minimum acceleration, and the maximum velocity into the first row of the table on the next page.

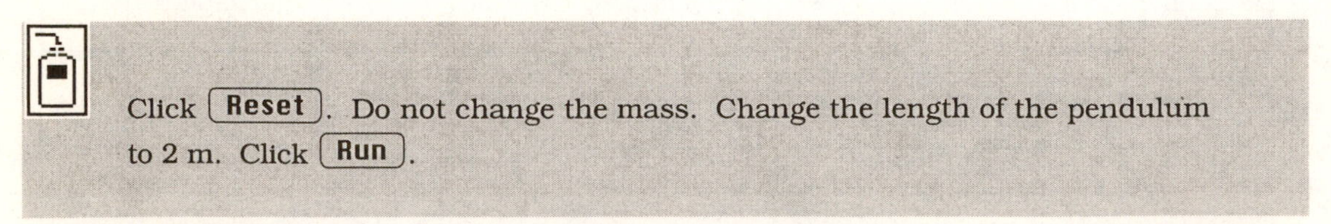

Click Reset. Do not change the mass. Change the length of the pendulum to 2 m. Click Run.

From the simulation data, record the maximum tension, the minimum tension, the period, the maximum acceleration, the minimum acceleration, and the maximum velocity into the second row of the table on the next page.

Click [Reset]. Do not change the mass. Change the length of the pendulum to 1 m. Click [Run].

From the simulation data, record the maximum tension, the minimum tension, the period, the maximum acceleration, the minimum acceleration, and the maximum velocity into the last row in the table below.

Varying Pendulum Length (10°)

Length	F-max	F-min	T	a-max	a-min	v-max
4						
2						
1						

What two things (in this table) does changing the length of the pendulum affect? Are you surprised that the tensions and accelerations are unaffected? If you are, read the following explanation carefully.

Explanation

The maximum acceleration occurs at maximum swing and is equal to $g\sin\theta_0$, which does not depend on length. The acceleration at the low point depends on v and L, both of which change. Consider increasing the length by a factor of four. This increases the velocity by a factor of 2, and since the acceleration = v^2/L, the changes cancel exactly and the acceleration and the tension at the low point are the same. The tension at the maximum swing depends on the initial displacement and not the length.

Now you will investigate how things do or do not change when the length of the pendulum is kept fixed, but the initial displacement is varied. First copy the data on a pendulum of length 2 meters (second row) from the last table into the first row of the table on this page.

Click [Reset]. Leave the mass unchanged and set the length to 2 m. Run three more times with the initial angles given in the table below. For each case fill the data into the table.

Varying Initial Displacement (L = 2m)

initial angle	F-max	F-min	T	a-max	a-min	v-max
10°						
30°						
45°						
60°						

Calculate the theoretical value for the period of this pendulum.

T = ________

We will define the percent difference by $\frac{T_{simulation} - T_{theory}}{T_{theory}}$. *What is the percent difference for 10°, 30°, 45°, and 60°?*

% difference for 10° = ________

% difference for 30° = ________

% difference for 45° = ________

% difference for 60° = ________

What happens to the maximum velocity as the initial displacement is increased?

What happens to the maximum tension in the rope as the initial displacement is increased?

What happens to the minimum tension in the rope as the initial displacement is increased?

Explanation

The maximum tension occurs (for all θ_o) at the lowest point and is given by

$$F = mg + \frac{mv^2}{L}.$$

The only variable here is the velocity at the low point, which depends on the initial potential energy of the bob. The greater the initial angle of release, the greater the potential energy and hence the greater the speed at the bottom. Your simulation data should support this. The minimum tension in the rope occurs at the end point in the swing, and since the velocity is zero at that point it is given by $mg\cos\theta_0$ which decreases as the initial displacement is increased.

Optional Challenge

You may have noticed that for large angles (like 60°), the acceleration is no longer a minimum at the lowest point. In fact it is a maximum at that point and the minimum value of acceleration occurs at some point other than the bottom or the endpoint. For those of you who want a real challenge (and you need calculus to do this) try to derive an expression for θ where the acceleration is a minimum. And if you want another challenge, prove that all pendulums with initial displacements of less than 41.4° behave like the 10° case (maximum acceleration at ends, minimum at the bottom) and all pendulums with initial displacements greater than 41.4° behave like the 60° case (maximum at the bottom, minimum elsewhere). No kidding!

Self-Test Questions for Simulation 33

The questions refer to the two sets of graphs for simple pendulums. As was done in the simulation, the velocity graphs include the sign, but the acceleration graphs are magnitude only. One set of graphs represents data for two pendulums with different initial displacements (but the same length), and the other set represents data for two pendulums with the same initial displacement (but different lengths).

SET A

velocity (including sign)

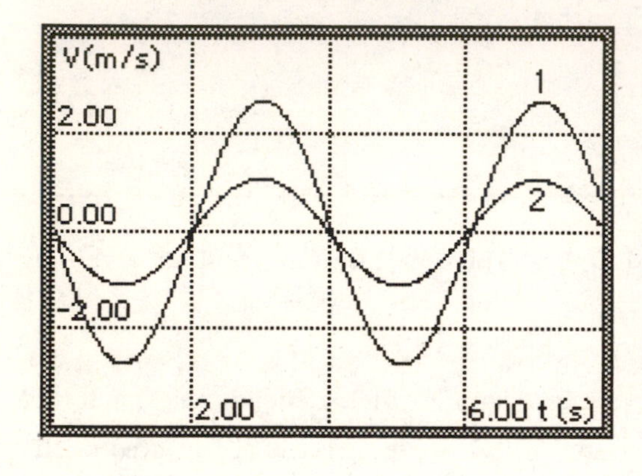

magnitude of the acceleration

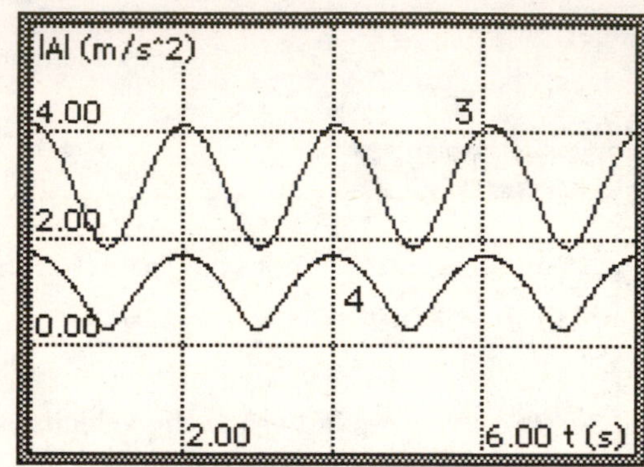

SET B

velocity (including sign)

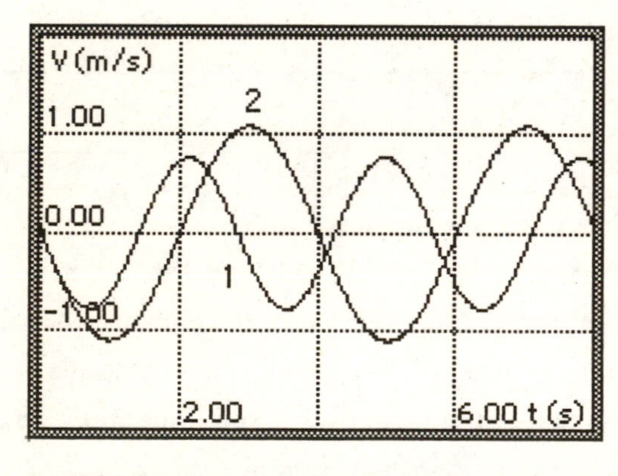

magnitude of the acceleration

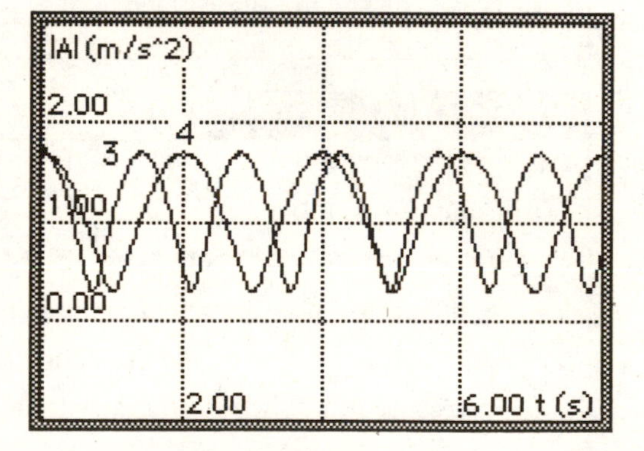

1. Which set represents data for two pendulums with different initial displacements?
2. Which set represents data for two pendulums with different lengths?
3. In set A, which acceleration curve (3 or 4) goes with curve 1?
4. In set B, which curve represents the pendulum with the largest maximum speed?
5. In set B, which curve represents the pendulum with the largest maximum acceleration?

SIMULATION 34 ROLLING DOWN HILL

Physics Review

In the previous problems that involved inclined planes, you always looked at blocks or crates moving up or down. These rectangular objects were only sliding and therefore you were able to analyze the problems as one-dimensional motion involving translation and not rotation. The situation changes when you place a ball on an inclined plane. The ball can slide down the hill (if the coefficient of friction is zero), and then the motion is purely translational and you can treat it the same way that you did for the blocks and crates. If the coefficient of friction is high enough, then the ball will roll down the hill and you must consider translational motion (i.e., the motion of the center of mass of the ball) and rotational motion (the motion of points on the ball about its center). There are several ways to approach this problem: you will use energy considerations. For the case of pure rolling motion (no slippage) we will assume that no energy is lost. Then the linear velocity of the center of the ball is related to the rotational velocity (of the ball about its center) as $v = \omega R$. Conservation of energy gives us:

$$mgh + \tfrac{1}{2}mv^2 + \tfrac{1}{2}I\omega^2 = constant$$

where the last term on the right-hand side is the rotational kinetic energy of the rolling object, which depends on the moment of inertia (I) of the object. You can probably find the formulas for the moment of inertia of various objects in your textbook, but here are some that you will be using in this simulation.

$I_{hoop} = MR^2$ $\qquad$ $I_{solid\ ball} = \tfrac{2}{5}MR^2$

Reference topics: rolling, moment of inertia, rotational motion

Simulation Details

In this simulation a hoop is ready to roll down a hill inclined at an angle of 25°. The mass of the hoop is adjustable with a slider, and the hoop can be "changed into" a solid ball, hollow ball, or cylinder by adjusting the moment of inertia factor. This factor is just the number that multiplies MR^2. For a hoop this number is 1.0.

To "turn the hoop into a sphere," type 0.4 into the factor box (or slide to the sphere). The radius of the hoop/sphere is fixed at 0.3 m. Meters for position, kinetic energy (translational and rotational), potential energy, acceleration (linear), velocity (linear of the center (v) and rotational (ω) about the center) of the hoop are included. You can adjust the coefficient of friction between the rolling object and the hill. To assure pure sliding, use μ = 0.0, and to assure pure rolling, use μ = 1.0. A point is chosen on the rim of the hoop so that you can visually see the rolling. The velocity of this point (point A) is also measured on a numerical meter. The simulation will stop when the object gets near the bottom of the hill.

Set the mass of the hoop to 0.5 kg and make sure that the moment of inertia factor is set to 1.0. Set the coefficient of friction to 0.0 to assure that the hoop will slide down the hill (like a hula-hoop on an icy slope). Click **Run** and observe the motion.

From the simulation data, fill in the following table.

Hoop Sliding

	v-linear	ω	KE-trans	KE-rot	PE
initial					
final					
acceleration =			time to get down the hill =		

Use the data that you just recorded and the [icon] to answer the following questions.

How far has the hoop gone?

Is energy conserved?

Click **Reset**. Change the coefficient of friction to 1.0 to assure that the hoop will roll without slippage down the hill. Click **Run** and observe the motion.

From the simulation data, fill in the following table.

Hoop Rolling

	v-linear	ω	KE-trans	KE-rot	PE
final					
acceleration =			time to get down the hill =		

Use the data that you just recorded and the [tape icon] to answer the following questions.

What is the ratio of the translational kinetic energy to the rotational kinetic energy?

Why did it take longer for the hoop to roll down the hill than it did for it to slide down the hill? Is this what you expected would happen?

How many revolutions does the hoop make as it rolls down the hill?

Explanation

For pure rolling without any slippage, the linear distance traveled must be equal to the number of revolutions times the circumference of the hoop. *Is it true in this case?* Recall that the radius of the hoop is 0.3 m.

Use the to go back and find a time where point A is at the top of the hoop (i.e., 180° from the contact with the hill). From the simulation data, record or calculate the following:

velocity of the point = ____________

linear velocity of the hoop = __________

What is the ratio of the velocity of point A to the linear velocity of the hoop?

Use the tape player to go back and find a time where point A is in contact with the hill. From the simulation data, record or calculate the following:

velocity of the point = _________

Explanation

Many books use the following diagram showing the vector superposition of translation and rotation to explain rolling. For pure rolling motion the velocity of the center of the ball is just the linear velocity v, the velocity of a point on top of the hoop is twice that and the velocity of a point in contact with the surface that the ball is rolling on is zero.

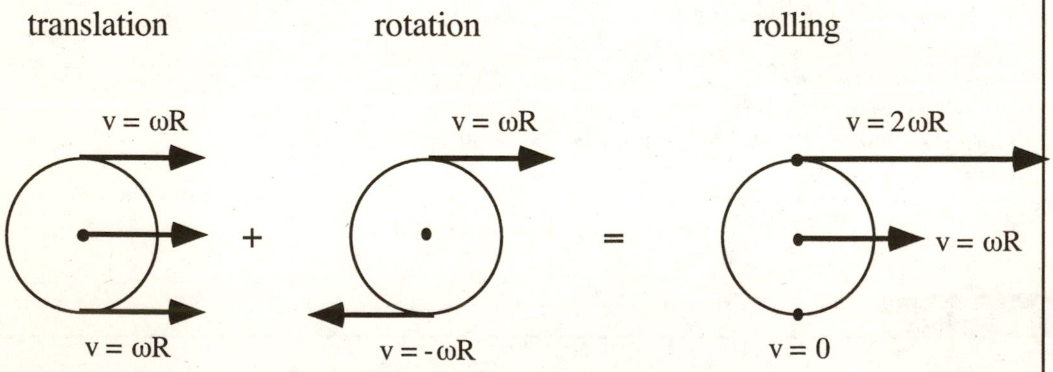

translation rotation rolling

$v = \omega R$ $v = \omega R$ $v = 2\omega R$

\+ = $v = \omega R$

$v = \omega R$ $v = -\omega R$ $v = 0$

Do your observations and the simulation data support this theory?

Click **Reset**. Keep the mass of the object the same, but change it into a solid spherical ball by changing the moment of inertia factor to 0.4. Set the coefficient of friction to 1.0 to assure that the ball will roll without slipping. Click **Run**.

From the simulation data, fill in the following table.

Solid Sphere Rolling

	v-linear	ω	KE-trans	KE-rot	PE
final					
acceleration =			time to get down the hill =		

Compare this data to that which you recorded on page 203.
Which took longer to roll down the hill, the hoop or the sphere? Why?

Which had a larger acceleration, the hoop or the sphere?

What is the ratio of the translational kinetic energy to the rotational kinetic energy for the sphere? How does this compare to the case of the hoop where the ratio was 1:1?

What is the ratio of the linear velocity to the rotational velocity?

Is this ratio equal to the radius of the sphere?

Keep the coefficient of friction set to 1.0 to assure rolling and vary the moment of inertia factor, running each time and recording the acceleration in the table below. You need not run all the way.

Acceleration Data

factor	0.1	0.2	0.4	0.6	0.8	1.0
acceleration						

Acceleration vs. Moment of Inertia Factor

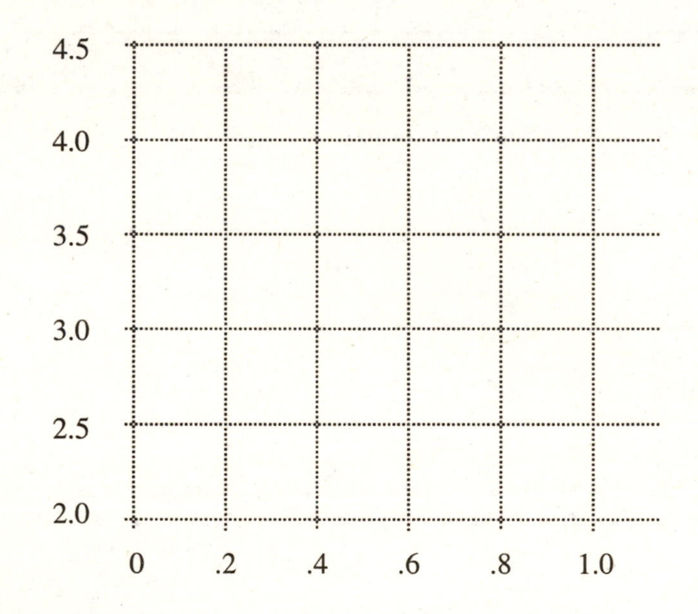

Explanation

It is possible to derive an equation for the acceleration of a rolling object with a moment of inertia, $I = f\,MR^2$. It is $a = g\sin\theta/(1+f)$. *Does your simulation data agree with this expression for acceleration?* (Optional: Derive this expression.)

Now you will investigate how high the coefficient of friction of the hill must be to assure pure rolling with no slippage for the hoop and the solid sphere. Assume a coefficient of friction of 0.2. Will the sphere roll or slip? Will the hoop roll or slip? Let's investigate this using the simulation data. There are three conditions that you can check to see if pure rolling is occurring. They are equivalent, that is, if one is satisfied, all will be and if one is not satisfied, none will be.

The three conditions are

1) $\dfrac{KE_{trans}}{KE_{rot}} = \dfrac{1}{f} = \dfrac{5}{2}$ for the sphere and $\dfrac{KE_{trans}}{KE_{rot}} = \dfrac{1}{f} = \dfrac{1}{1}$ for the hoop

2) $\dfrac{v}{v_{rot}} = R$

3) $N = \dfrac{linear\ distance}{2\pi R}$ where N is the number of revolutions made

Click **Reset**. Set the coefficient of friction to 0.2 and the moment of inertia factor to 0.4 for a solid sphere. Click **Run** and observe the motion until the sphere gets to the bottom.

Check each of the three conditions for rolling and decide whether it is slipping or not.

1)

2)

3)

Click **Reset**. Keep the coefficient of friction at 0.2 and set the moment of inertia factor to 1.0 for a hoop. Click **Run** and observe the motion.

Check each of the three conditions for rolling and decide whether it is slipping or not.

1)

2)

3)

Conclusion: When the coefficient of friction is 0.2, the ______________________ will roll, but the ______________________ will slip as it rolls.

Optional: Derive the three conditions for pure rolling given at the top of page 207.
Hint: Use energy conservation for #1.

Self-Test Questions for Simulation 34

These statements all pertain to spheres or hoops rolling down an inclined plane. True or false?

1. If the coefficient of friction is zero, solid spheres and hoops will reach the bottom of the hill at the same time.

2. If the coefficient of friction is 1.0, spheres and hoops will both roll but will not arrive at the bottom at the same time.

3. For pure rolling motion, the final velocity of the hoop is equal to the final velocity of the solid sphere.

4. The velocity of a point in contact with the hill is zero for pure rolling motion.

5. If the velocity of a point at the very top of the hoop is 10 m/s and the linear velocity of the hoop at that instant is 5 m/s, the ball must not be rolling.

SIMULATION 35 ODD-SHAPED ROTATING PROJECTILE

Physics Review

Many kinds of motion cannot be analyzed using one single set of concepts. In most of the simulations you have done, each tended to focus primarily on one topic, like collisions *or* freefall *or* motion with friction *or* rotation. It is necessary to focus on and solve these "single concept" problems first before attempting to solve or understand situations that combine two or more of them. This problem will be the first of four problems that involve putting together several concepts. This problem involves projectile motion, rotational motion (including torque, angular momentum, and angular velocity), and several forms of mechanical energy. All of these topics have been discussed in the **Physics Review** sections of previous simulations (except angular momentum) so you can go back to some of those if you need review. The angular momentum (*L*) is the rotational analog to linear momentum and is given by the formula shown below.

Formulas you should be familiar with:

$v_x = v_{ox}$ $\qquad v_y = v_{oy} - gt$

$KE_{rot} = \frac{1}{2} I\omega^2$ $\qquad \tau = I\alpha$ $\qquad \omega = \omega_o + \alpha t$ $\qquad L = I\omega$

$KE = \frac{1}{2} mv^2$ $\qquad PE = mgh$

Reference topics: projectile motion, rotation, torque, angular momentum, energy

Simulation Details

In this simulation an odd shaped boomerang is thrown into the air. It is an idealized problem in that there is no air resistance. As the boomerang travels through the air, the position of its center of mass is tracked. A point near the edge of the boomerang (point A) is also tracked through the flight. It would seem that both the center of mass and point A would follow a parabolic path characteristic for projectiles. However, to make this motion more interesting, the boomerang is initially rotating at an unknown angular speed and there is an adjustable torque that can be applied to it. Output graphs show position (x and y) of the center of mass of the boomerang and point A. These curves are not labeled and it will be up to you to decide which is which. Another graph has four curves (you may only see three if two overlap), one representing each of v_x, v_y, $|v|$, and ω of the boomerang (also not labeled). The simulation will stop when the boomerang has returned to (near) its original y position.

Set the torque to 0.0 Nm. Click Run.

Which tracked path represents the trajectory of the center of mass of the boomerang and which represents the trajectory of point A?

Which curve represents the x position and which represents the y position of the center of mass of the boomerang?

Which curve represents the x position and which represents the y position of point A?

Which velocity curve for the boomerang represents v_y?

Which velocity curve for the boomerang represents $|v|$?

Which velocity curve for the boomerang represents v_x?

Which velocity curve for the boomerang represents ω?

From the simulation data, record or calculate the following:

the initial angular velocity of the boomerang = ___________

the final angular velocity of the boomerang = ___________

rotational *KE* at 2 sec = ___________

translational *KE* at 2 sec = _________

PE at 2 sec = ___________

Click **Reset** and change the torque to –5.0 Nm. Click **Run** and observe the motion.

Which trajectory has changed, the path of point A or the path of the center of mass?
Explain why.

Which of the velocity curves have changed?
Explain why.

Which of the position curves have not changed?
Explain why.

What is happening to the boomerang as it moves along? In particular, in what direction is it rotating? Explain why.

Do you still agree with the curve assignments that you made on the previous page? (Now you should be able to see the blue velocity curve representing v_x. It was hidden before by the orange curve.)

From the simulation data, record or calculate the following:

the initial angular velocity of the boomerang = __________

the final angular velocity of the boomerang = __________

rotational *KE* at 2 sec = __________

translational *KE* at 2 sec = __________

PE at 2 sec = __________

Compare these values to the simulation data (for no torque) that you recorded on the previous page.

Which of the above quantities changed?
Explain why.

Which did not change?
Explain why.

Click [Reset] and change the torque to 2.0 Nm. Click [Run] and observe the motion.

From the simulation data, record or calculate the following:

the initial angular velocity of the boomerang = ___________

the final angular velocity of the boomerang = ___________

rotational *KE* at 2 sec = ___________

translational *KE* at 2 sec = ________

PE at 2 sec = ___________

Compare these values to the simulation data that you recorded for no torque (page 211) and for a torque of –5 Nm (page 212).

Which of the above quantities changed?
Explain why.

Which did not change?
Explain why.

Calculate the torque needed so that the angular speed of the boomerang will be zero when it is at its final position.

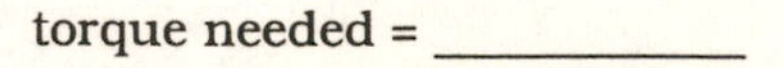
torque needed = ___________

Click Reset. Set the torque to this value (by typing it into the box) and click Run to check your calculation.

Is the angular speed of the boomerang close to zero when the simulation stops?

Self-Test Questions for Simulation 35

The following questions apply to a boomerang just like the one in the simulation. It has a fixed initial speed and a fixed initial angular speed. True or false?

1. The velocity of the center of mass of the boomerang does not depend on how the boomerang rotates.
2. The rotational kinetic energy of the boomerang is negative if the torque on it is negative.
3. The potential energy of the boomerang depends on the torque.
4. As the rotational speed of the boomerang increases, so does its total mechanical energy.
5. The time for the flight does not depend on the torque.

SIMULATION 36 MYSTERY MOTION 1

Simulation Details

The mass of a box is 0.5 kg, and it is sliding on a horizontal surface. You are looking down on it from above. The surface that it moves on may not be the same in all areas. You must figure out the "mystery of the motion," in particular, what forces are acting and what kind they are. One of the forces is a force of friction. The simulation will stop when 2 seconds have elapsed. This simulation is more free form, and you will not be given as much direction. There are several methods that can be used. Topics included in this simulation include: forces, work, friction and energy. You should not attempt to solve the "mystery" unless you have covered these topics.

Determine the magnitude and direction of the first force that acts on the box and how much work it does.

magnitude = ______________

direction = ________________

work done = ______________

Determine the magnitude and direction of the second force that acts on the box and how much work it does.

magnitude = ______________

direction = ________________

work done = ______________

Determine the magnitude and direction of the third force that acts on the box and how much work it does.

magnitude = ______________

direction (angle wrt positive x direction) = ______________

work done = ______________

Determine the approximate coefficient of friction for the frictional force.

μ_k = __________

This is a plot of the kinetic energy of the box as a function of time. Mark the region where each force acts. Use different colors if you have them handy.

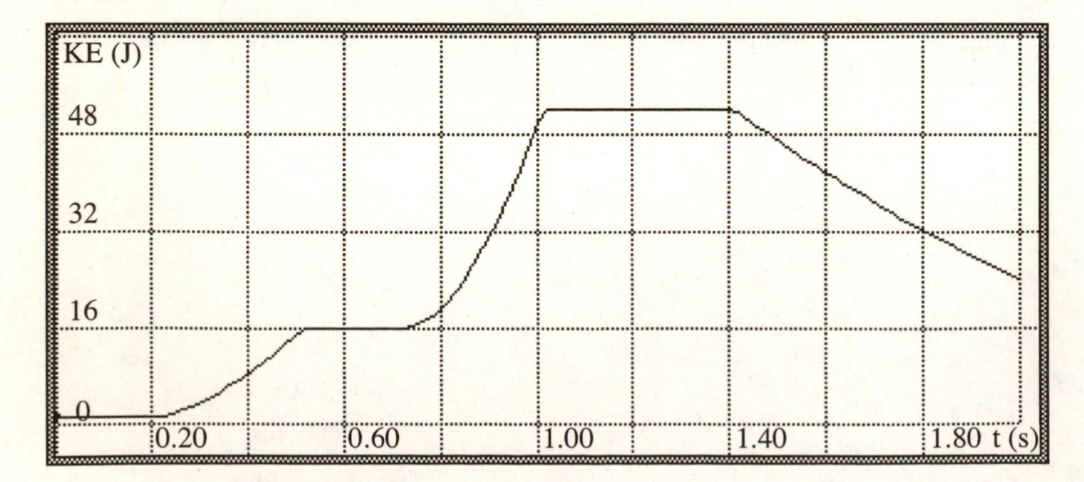

SIMULATION 37 MYSTERY MOTION 2

Simulation Details

Three balls (black, gray, and striped) are all headed initially straight toward a wall. All three collide with the wall. Output meters give simulation data for the momentum (p_x, p_y, and $|p|$) and kinetic energy of each ball. It is your goal to figure out several "mysteries of the motion". You may choose to do these in order or skip around. I suggest that you observe the motion several times before attempting to answer any of the questions. This simulation will automatically stop at some point.

Topics included in this simulation include: collisions, and kinetic and potential energy. You should not attempt to solve the "mystery" unless you have covered these topics.

Determine the initial velocity of each ball.

initial velocity of the black ball = ________________

initial velocity of the gray ball = ________________

initial velocity of the striped ball = ________________

Determine the mass of each ball.

mass of the black ball = ________________

mass of the gray ball = ________________

mass of the striped ball = ________________

Determine where this motion is occurring (*e.g.*, the Earth, the Moon, Jupiter).

Determine which collision(s) are inelastic and which are elastic.

The gray ball loses some of its energy when it collides with the wall and all of its energy when it collides with the floor. *Why?*

Determine the average force exerted by the wall on each ball during the collision.

force on the black ball = ________________

force on the gray ball = ________________

force on the striped ball = ________________

SIMULATION 38 COMBINATION MOTION 1

Simulation Details

A block of mass 3 kg is pressed up against a spring so that the spring is compressed 1 meter from its equilibrium position. The block is resting on a horizontal surface whose coefficient of friction is not zero. The system is released, and once the block passes through the point $x = 0$, it is free from the spring. It slides on the horizontal surface until it collides with another block. This simulation involves many of the concepts that you have covered in previous simulations, and you should not attempt it unless you have covered springs and Hooke's law, kinetic energy, friction, and collisions. This simulation is similar to the previous two in that you will not be given any explicit instructions and it does not have self-test questions. Note that there may be several ways to approach this problem. Run the simulation all the way through until it stops, and then proceed with the calculations and questions.

Determine the coefficient of friction between the blocks and the surface.

μ_k = ________

Determine the spring constant.

k = ________

Determine the mass of block 2.

mass = ____________

Determine if the collision is elastic or not.

SIMULATION 39 COMBINATION MOTION 2

Simulation Details

In this simulation a boulder is set to roll down a hill and subsequently up and over two other hills. You can adjust the mass of the boulder and its initial speed. If it makes it over the second hill, it will collide with a box.

Does the mass of the boulder affect whether it makes it up and over the first hill? Why or why not?

The second hill is 3.4 meters higher than the first hill. Calculate the initial velocity needed to get the boulder to just make it over the top of the second hill.

initial velocity of the boulder = ________________

Set the mass of the boulder to the smallest value and its speed to the maximum value. Run until the boulder collides with the box. Use the simulation data to find the mass of the box.

mass of the box = ________________

Determine if the collision is elastic.

The mass and speed of the boulder will determine the direction of the boulder after the collision. Explain this.

SIMULATION 40 DOUBLE PENDULUM

Simulation Details

Here we have created a simulation of a double pendulum, one simple pendulum suspended from another (in this case we have used inextensible rods instead of ropes). There is no air resistance. This problem is beyond the scope of an introductory physics class, but is typically solved in a sophomore level mechanics class. Students in such a course can solve this problem using more advanced mathematical methods. The result is a rather complicated set of equations representing the motion. Looking at the equations does not give much insight into the motion of either pendulum bob or answer questions like "How does the velocity of each vary with time?" or "Is there any regularity to the motion?" Even though most of you do not have the mathematical background to solve this problem, we have included it for two reasons:

1) To show you some more interesting things that you can do in *Interactive Physics.*
2) To show you one of the many exciting things that may be in store for you to study if you go on to take more physics classes.

So, go open the file, if you have not already done so, run it and observe for a few seconds. Try to answer the following questions.

Do you think that the motion is periodic?

Is either pendulum undergoing simple harmonic motion?

Does changing either mass affect the motion?

What does the path of the bottom bob look like in the reference frame of the top bob?

Do you think that energy is conserved in this problem?

Feel free to edit this file and investigate further if you have the full *Interactive Physics* Program .

Simulation 1

1. A car with positive velocity that is increasing in magnitude must have a positive acceleration. **T**
2. A car that has a negative acceleration must be slowing down. **F**
3. A car that has a velocity and acceleration in opposite directions is slowing down. **T**
4. A car that has a constant velocity must have zero acceleration. **T**
5. If a car reverses direction, there must be an acceleration opposite to its original velocity. **T**

Simulation 2

1. In which graph(s), if any, does the car come to a stop (even momentarily)? **2, 3, and 4**
2. In which graph(s), if any, does the car change directions? **3 and 4**
3. Which graph represents the car whose acceleration has the largest magnitude? **4**
4. Which graph represents the car with the largest initial speed? **3**
5. In which graph is the car moving at a constant speed? **1**

Simulation 3

1. Is the acceleration of the car positive or negative? **positive**
2. Is the initial velocity of the car positive or negative? **negative**
3. Which way is the car moving for the first 5 seconds—to the right or the left? **left**
4. At what time does the car come momentarily to a stop? **at about 5 seconds**
5. Does the car ever return to its initial starting position? If so, at what time? **yes, at about 10 seconds**

Simulation 4

1. The velocity graph of car A is a horizontal line. **T**
2. The velocity graph of car B is not a straight line. **F**
3. Car B can never catch up with car A. **F**
4. Car C can never catch up with car A. **T**
5. During any time interval, car B goes the farthest. **F**

Simulation 5

1. The velocity of the boat with respect to the shore depends on the velocity of the river. **T**
2. It is not possible for the boat to arrive at a point directly across the river. **F**
3. The motion of the boat is different for different observers. **T**
4. The time that it takes the boat to cross the river is the same regardless of who is measuring it (e.g., the person in the boat or the person at the shore). **T**
5. The total distance that the boat travels in crossing the river depends on the speed of the river and the heading of the boat. **T**

Simulation 6

1. Which is correct, $\vec{C} = \vec{A} + \vec{B}$ or $\vec{B} = \vec{A} + \vec{C}$? $\mathbf{\vec{B} = \vec{A} + \vec{C}}$
2. What is the speed of the wind? **about 100 m/s**
3. What is the speed of the plane with respect to the ground? **about 173 m/s**
4. How long does it take to get from point O to the airport? **about 5.8 seconds**
5. What direction is the wind coming from? **from the southeast**

Simulation 7

1. The velocity of the ball is zero at the highest point. **T**
2. When the ball returns to its original position, its velocity has the same magnitude as the initial velocity. **T**
3. For the same initial speed, the time it takes for the ball to reach its maximum height on the Earth is 6 times longer than the time that it takes the ball to reach its maximum height on the Moon. **F**
4. On both the Earth and Moon the ball's acceleration is zero at its maximum height. **F**
5. For the same initial speed, the ball goes 6 times higher on the Moon than it does on the Earth. **T**

Simulation 8

1. Which height curve represents the ball launched with the greatest initial speed? **1**
2. About how long does it take the ball represented by curve 2 to complete its trip and return to the ground? **about 1.8 seconds**
3. Approximately how high does the ball represented by curve 1 go? **about 20 meters**
4. What was the approximate initial speed of the ball represented by curve 1?
   about 20 m/s Note: There are two ways to arrive at this conclusion. Read the time to the top off the curve 1 graph, which is about 2 seconds, and use
   $v = v_o - gt$
   or note that curve 3 starts at a value of about 2/3 the *y* axis mark of 32 m/s.
5. At approximately what other time is the height of the ball represented by curve 1 the same as it was at 1 second? **approximately 3 seconds**

Simulation 9

1. The velocity at the top of the flight is zero. **F**
2. For a given initial speed, a larger angle results in a longer flight time. **T**
3. For a given initial speed, a larger angle results in a longer range. **F**
4. The x component of the velocity is constant throughout the entire motion. **T**
5. The speed of the projectile decreases on the way up and increases on the way down. **T**

Simulation 10

1. Which six curves (one from each graph) correspond to the motion with air resistance?
   2,3,5,7,10,12
2. On the graph of the y component of velocity, which curve represents the ball that takes longer to reach its maximum height? **6 (no air resistance)**
3. Which ball has fallen (y position) farther after 2.5 seconds have elapsed?
   10 (air resistance)
4. Which ball will have fallen (y position) farther at 3.5 seconds?
   9 (no air resistance)
5. Which ball has gone farther (x position) in 2 seconds? **11 (no air resistance)**

Simulation 11

1. The time of flight of the ball is the same for all observers. **T**
2. The ball always lands in the sports car. **T**
3. The velocity of the ball at the top of the flight is the same for all observers. **F**
4. Only the people in the sports car can say that $d = vt$. **F**
5. The path the ball takes depends on the observer and her velocity with respect to the sports car. **T**

Simulation 12

1. Label the following free body diagram for the puck with the forces 1, 2, and 3.

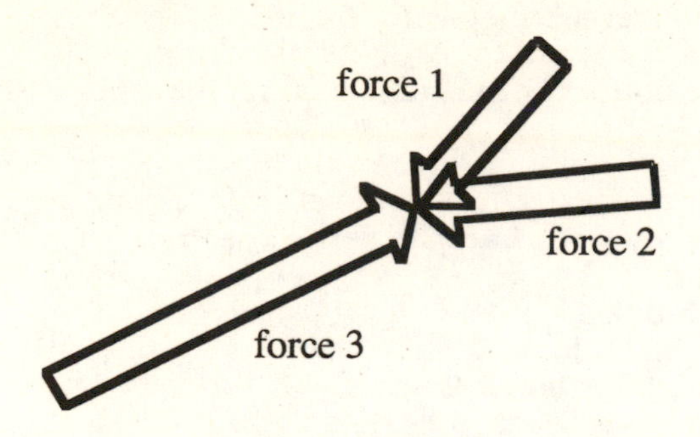

2. Can you tell from this information whether or not the puck is moving? **No, you can only tell that it is in equilibrium.**
3. Is the puck accelerating? **No, since the net force on it is zero, it is not accelerating.**
4. Which force has the largest magnitude? **force 3**
5. Is there any other value for force 1 that will result in equilibrium? **No, given force 2 and 3, force 1 is uniquely determined.**

Simulation 13

1. Which block has the smaller mass? **block 2**
2. Which block loses more energy? **block 1**
3. Which block requires more work to stop it? **block 1**
4. Which block's acceleration has the largest magnitude? **block 2**
5. Which one takes longer to stop? **block 1**

Simulation 14

1. The kinetic energy of the crate is always increasing as long as the external force is being applied. **F**
2. The work done by the external force is always equal to the work done by the frictional force. **F**
3. When the external force has a fixed direction and $0^\circ < \theta < 90^\circ$, increasing the magnitude of the force will always result in a larger acceleration. **T**
4. An angle of 45° results in a smaller force of friction than an angle of 36°. **T**
5. An angle of 30° results in a larger normal force than an angle of –30°. **F**

Simulation 15

1. If the mass of the block on the table is larger than the mass of the hanging block and there is no friction, the system will not accelerate. **F**
2. The tension in the rope will never be larger than the weight of the hanging block. **T**
3. The tension in the rope does not depend on the mass of the block on the surface. **F**
4. It is not possible for the hanging block to move upward. **T**
5. For fixed values of the two masses, the acceleration will be the smallest when the coefficient of friction is the largest. **T**

Simulation 16

1. The distance a block will travel up the hill depends on the initial speed of the block. **T**
2. If the mass of the block is larger, the distance traveled up the plane will be less. **F**
3. The smaller the mass, the less time it takes to travel 1 meter up the hill. **F**
4. The net force on the block is never zero. **T**
5. The speed of the block when it returns to the bottom is the same as its initial speed. **T**

Simulation 17

1. Which block had the greater initial speed? **They both had the same speed.**
2. Which block goes farther up the hill? **block 1**
3. Which block takes longer to go up the hill? **block 1**
4. Which block takes longer to come down the hill? **block 2**
5. Which block experienced a greater force of friction? **block 2**

Simulation 18

1. Which block lost more mechanical energy? **block B**
2. What percentage of the original energy is lost by block A? **about 40%**
3. Which block has more potential energy at $t = 2$ seconds? **block A**
4. Which block reaches its maximum height first? **block B**
5. Which block experiences the largest frictional force? **block B**

Simulation 19

1. During what time interval is the coefficient of friction the smallest? **1–2 seconds**
2. During what time interval is the coefficient of friction the largest? **$t > 2$ seconds**
3. Is there a time interval when the net force on the block is zero? **yes, $t < 1$ second**
4. During what time interval (if any) is the coefficient of friction equal to the tangent of the angle of inclination of the hill? **$t < 1$ second**
5. During what time interval (if any) is the coefficient of friction equal to zero? **from 1–2 seconds. For no friction, the acceleration would be 9.8*sin (21°), which is about 3.5 m/s^2, which is approximately equal to the slope of the curve during that time.**

Simulation 20

1. The tension T_1 can never be greater than F. **T**
2. The tension T_2 can never be greater than T_1. **T**
3. If the 2 kg and 1 kg crates are exchanged, the acceleration of the system will decrease. **F**
4. The tension in each rope depends on the coefficient of friction. **F**
5. The tension in each rope depends on the force F. **T**

Simulation 21

1. The acceleration of the crates is less than 2 m/s^2. **T**
2. Switching the first and second crate will change the acceleration. **F**
3. Switching the first and second crates will change the tensions in both of the ropes between the three crates. **F**
4. The change in kinetic energy is the same for each crate. **F**
5. The change in the potential energy is the same for each crate. **F**

Simulation 22

1. All collisions conserve momentum. **T**
2. All collisions conserve kinetic energy. **F**
3. A completely inelastic collision results in a maximum amount of kinetic energy lost. **T**
4. For any perfectly elastic collision between two objects of equal mass, they will always exchange velocities. **T**
5. Completely inelastic collisions (where the objects stick together) typically involve energy losses of only about 10%. **F**

Simulation 23

1. The initial kinetic energy of the ball depends on the launch angle. **F**
2. The potential energy is a maximum at the top of the flight. **T**
3. If the ball is launched at 30°, with no air resistance, the kinetic energy at the top is zero. **F**
4. The kinetic energy of the ball launched with no air resistance is always positive. **T**
5. When a ball is launched at 30° with air resistance, its kinetic energy decreases, then increases, then approaches a constant value. **T**

Simulation 24

1. Is the initial kinetic energy of the ball in A greater than, less than, or equal to the initial kinetic energy in B? **They are equal because they have the same initial velocity and the same mass.**
2. Were the collisions (A and B) elastic? **no**
3. Which graph represents the case with the lower ceiling? **B**
4. Which graph represents the ball that lost more energy overall? **B**
5. Which graph represents the ball with the smaller speed after the collision? **A**

Simulation 25

1. Which graphs represent balls with a speed of 20 m/s? **2 and 3**
2. Which graph represents a ball moving in a circle with the highest frequency? **3**
3. Which two graphs show the balls moving with the same period? **2 and 4**
4. What is the approximate ratio of the radius of the motion of ball 2 to ball 4? **2:1**
5. Which ball is moving in a circle in a direction opposite to the other three? **3**

Simulation 26

1. When the angular acceleration of the rod is zero, its angular speed is constant. **T**
2. The angular velocity of one end of the rod is the same as the angular velocity of the other end of the rod. **T**
3. The linear velocity of a point on the rod depends on its distance from the center of the rod. **T**
4. If the angular speed of a rotating rod increases from zero to 5 rad/s in 0.4 seconds, what is its angular acceleration? **12.5 rad/s^2**
5. What will be the angular displacement of that rod at t = 0.4 seconds? **1 rad**

Simulation 27

1. What is the approximate angular acceleration of the door represented by curve 1? **6 rad/s^2**
2. Approximately how many revolutions does the door represented by curve 2 make in the last second? **about 1**
3. If you keep the magnitude and direction of the force fixed, where must you apply it to achieve the data represented by curve 2? **at 2/3 L from the center**
4. If you keep the magnitude and point of application of the force fixed, at what angle must the force be applied to achieve the data represented by curve 2? **about 42°**
5. If you keep the direction of the force and its point of application fixed, what magnitude must the force have to achieve the data represented by curve 2? **2/3 of F**

Simulation 28

1. The sum of the forces can be zero even if the sum of the torques is not zero. **T**
2. The torque exerted by any block about the center of the meter stick will be the same as the torque exerted by the same block about the end of the meter stick. **F**
3. The force exerted on the meter stick by the support is not zero. **T**
4. The net torque about the right end is the same as the net torque about the left end. **T**
5. Using our sign convention, a block hung to the left of the center will result in a negative torque about the center. **F**

Simulation 29

1. A constant torque will produce a constant angular acceleration. **T**
2. A positive torque will produce an increasing angular acceleration. **F**
3. A zero torque will produce a zero angular acceleration. **T**
4. A negative torque, applied to a wheel initially at rest, will produce a centripetal acceleration that is increasing. **T**
5. A constant centripetal acceleration is caused by a constant torque. **F**

Simulation 30

1. Is the torque positive or negative during the first 8 seconds? **negative**
2. How many times does the teacup reverse direction? **once**
3. At what time(s) does the teacup reverse direction? **at about 21 seconds**
4. Is the speed of the teacup increasing or decreasing during the first 8 seconds? **increasing**
5. During what time period is the speed of the cart increasing at the fastest rate? **after 21 seconds**

Simulation 31

1. Which velocity curve goes with acceleration curve 1? **3**
2. Which acceleration curve represents the block with the smaller mass? **1**
3. What is the approximate ratio of the masses? **2:1, from** $m_1 a_{max1} = kA = m_2 a_{max2}$
4. Which block has the greater speed at 3 seconds? **3**
5. Which block experiences the greater maximum force? **They both experience the same maximum force.**

Simulation 32

1. About how long does it take for the block in graph B to stop? **about 3.5 seconds**
2. Which spring has a greater spring constant, or are they the same? **They are the same.**
3. Which numbered curve represents the kinetic energy of a block? **3**
4. Which graph represents the block with a greater mass? **B**
5. Is there ever a time where the kinetic energy of a block is equal to the potential energy of its spring? **Yes, several times, when the curves (2 and 3) cross.**

Simulation 33

1. Which set represents data for two pendulums with different initial displacements? **A**
2. Which set represents data for two pendulums with different lengths? **B**
3. In set A, which acceleration curve (3 or 4) goes with curve 1? **curve 3**
4. In set B, which curve represents the pendulum with the largest maximum speed? **curve 2**
5. In set B, which curve represents the pendulum with the largest maximum acceleration? **They have the same maximum acceleration.**

Simulation 34

1. If the coefficient of friction is zero, solid spheres and hoops will reach the bottom of the hill at the same time. **T**
2. If the coefficient of friction is 1.0, spheres and hoops will both roll but will not arrive at the bottom at the same time. **T**
3. For purely rolling motion, the final velocity of the hoop is equal to the final velocity of the solid sphere. **F**
4. The velocity of a point in contact with the hill is zero for pure rolling motion. **T**
5. If the velocity of a point at the very top of the hoop is 10 m/s and the linear velocity of the hoop at that instant is 5 m/s, the ball must not be rolling. **F**

Simulation 35

1. The velocity of the center of mass of the boomerang does not depend on how the boomerang rotates. **T**
2. The rotational kinetic energy of the boomerang is negative if the torque on it is negative. **F**
3. The potential energy of the boomerang depends on the torque. **F**
4. As the rotational speed of the boomerang increases, so does its total mechanical energy. **T**
5. The time for the flight does not depend on the torque. **T**

Simulation 36

The first force has a magnitude of 10 N (from $F = ma$), acts in the x direction for 0.3 s, and does about 15 J of work (from the change in kinetic energy).

The second force has a magnitude of 20 N (from $F = ma$) in the y direction, acting for 0.3 s too, and it does about 36 J of work (from the change in kinetic energy).

The third force is frictional. Its magnitude is 3.92 N, and its direction is opposite to the velocity of the ball (236°). It does about −28 J of work. The coefficient of friction is 0.8.

Simulation 37

From KE and momentum data: you divide the KE by the momentum and that gives $1/2\ v$. Giving:

the velocity of the black ball = 3 m/s
the velocity of the gray ball = 3 m/s
the velocity of the striped ball = 6 m/s

From $p = mv$: since you know the initial momentum and velocity for each ball, you can calculate the mass. Giving:

the mass of the black ball = 1 kg
the mass of the gray ball = 4 kg
the mass of the striped ball = 2 kg

The motion takes place on the Moon. Study the y component of the velocity data.

The collision of the black ball is elastic and the other collisions are inelastic: Look at changes in kinetic energy.

Average force and change in momentum are related by $\Delta p = F\Delta t$, therefore:

the average force exerted on the black ball is about 300 N
the average force exerted on the gray ball is about 600 N
the average force exerted on the striped ball is about 900 N

Simulation 38

The coefficient of friction is 0.1. Look at the acceleration of the block after it is free from the spring.

The spring constant is 30 N/m. The initial potential energy of the spring minus the work done by the frictional force is equal to the kinetic energy of the block.

The mass of block 2 is 6 kg as momentum is conserved in the collision.

The collision is not elastic because the kinetic energy after the collision is less than the kinetic energy before the collision.

Simulation 39

Use the increase in potential energy needed to calculate the initial kinetic energy. Mass doesn't matter, and the velocity needed is about 8.2 m/s.

Conservation of momentum in the collision gives a mass of about 0.5 kg for the box.

If the momentum of the boulder is small, then it will bounce off the block and move back up the hill. If the momentum is large, then it will knock the box off to the right and continue in that direction.

Kinetic energy is conserved so the collision is elastic.

Sim #	Reference Topics	Fishbane/ Gasiorowicz/ Thornton Physics 3e Calculus	Giancoli PSE 3e Calculus	Giancoli Physics 6e Algebra	Hobson Phys. C&C 3e Algebra	Walker Physics 2e Algebra	Wilson/Buffa Coll. Phys. 5e Algebra
1	one-dimensional motion, velocity, acceleration, kinematics	2	2	2	3	2	2
2	one-dimensional motion, velocity, acceleration, kinematics	2	2	2	3	2	2
3	one-dimensional motion, constant acceleration	2	2	2	3	2	2
4	one-dimensional motion, constant acceleration	2	2	2	3	2	2
5	vectors, vector addition, relative velocity	2,3	2,3	2,3	3	2,3,4	2,3
6	vectors, vector addition, relative velocity	2,3	2,3	2,3	3	2,3,4	2,3
7	free fall, falling bodies	2	2	2	3	2	2
8	free fall, falling bodies	2	2	2	3	2	2
9	Projectiles, projectile motion, vectors	2,3	2,3	2,3	3	2,3,4	2,3
10	air resistance, terminal velocity	2,3,5	2,3,5	2,3,5	3,4	2,3,4,6	2,3,4
11	relative velocity, reference frames, projectile motion	2,3	2,3	2,3	3	2,3,4	2,3
12	Newton's laws, forces, vectors	2,3,4	2,3,4	2,3,4	3,4	2,3,4,5	2,3,4
13	friction, normal force, work, kinetic energy	2,4,5,6,7	2,4,5,7,8	2,4,5,6	3,4,6	2,5,6,7,8	2,4,5
14	friction, normal force, work, kinetic energy	2,4,5,6,7	2,4,5,7,8	2,4,5,6	3,4,6	2,5,6,7,8	2,4,5
15	Atwood machine, pulleys, friction	2,4,5,6,7,9	2,4,5,7,8,10	2,4,5,6,8	3,4,5,6	2,5,6,7,8,10	2,4,5,7

Cross Reference to Texts (chapter number)

Sim #	Reference Topics	Fishbane/ Gasiorowicz/ Thornton Physics 3e Calculus	Giancoli PSE 3e Calculus	Giancoli Physics 6e Algebra	Hobson Phys. C&C 3e Algebra	Walker Physics 2e Algebra	Wilson/Buffa Coll. Phys. 5e Algebra
16	inclined planes, constant acceleration	2,3,4	2,3,4	2,3,4	3,4	2,3,4,5	2,3,4
17	inclined planes, friction	2,4,5	2,4,5	2,4,5	3,4	2,5,6	2,4
18	inclined planes, friction, energy	2,4,5,6,7	2,4,5,7,8	2,4,5,6	3,4,6	2,5,6,7,8	2,4,5
19	inclined planes, friction,	2,4,5	2,4,5	2,4,5	3,4	2,5,6	2,4
20	Newton's laws, friction	4,5	4,5	4,5	4	5,6	4
21	Newton's laws, friction, work/energy theorem	4,5	4,5	4,5	4	5,6	4
22	collisions, elastic collisions, inelastic collisions, momentum	6,7,8	7,8,9	6,7	4,6	7,8,9	5,6
23	energy, kinetic energy, gravitational potential energy	6,7	7,8	6	6	7,8	5
24	kinetic energy, collisions, gravitational potential energy	4,5,6,7	4,5,7,8	4,5,6	4,6	5,6,7,8	4,5,6
25	circular motion, period, frequency	2,3,9,10	2,3,10,11	2,3,8	3,5	2,3,4,10,11	2,3,7,8
26	circular motion, angular velocity, angular acceleration	9,10	10,11	8	5	10,11	7,8
27	angular acceleration, torque	9,10	10,11	8	5	10,11	7,8
28	equilibrium, torque	9,10,11	10,11,12	8,9	5,4	10,11	7,8
29	angular acceleration, centripetal acceleration, circular motion, torque	3,9,10	3,10,11	3,8	3,5	3,4,10,11	3,7,8
30	angular acceleration, centripetal acceleration, circular motion, torque	3,9,10	3,10,11	3,8	3,5	3,4,10,11	3,7,8

Sim #	Reference Topics	Fishbane/ Gasiorowicz/ Thornton Physics 3e Calculus	Giancoli PSE 3e Calculus	Giancoli Physics 6e Algebra	Hobson Phys, C&C 3e Algebra	Walker Physics 2e Algebra	Wilson/Buffa Coll. Phys. 5e Algebra
31	simple harmonic motion, period, amplitude, Hooke's law	4,6,7,13	4,7,8,14	4,6,11	4,6	5,7,8,13	4,5,13
32	simple harmonic motion, energy, friction	4,5,6,7,13	4,5,7,8,14	4,5,6,11	4,6	5,6,7,8,13	4,5,13
33	simple harmonic motion, simple pendulum	4,6,7,13	4,7,8,14	4,6,11	4,6	5,7,8,13	4,5,13
34	rolling, moment of inertia, rotational motion	9,10	10,11	8	5	10,11	7,8
35	projectile motion, rotation, torque, angular momentum, energy	3,4,5,6, 7,9,10	3,4,5,7,8, 10,11	3,4,5,6,8	3,4,5,6	3,4,5,6,7, 8,10,11	3,4,5,7,8
36	forces, friction, work, energy	2,3,4,5	2,3,4,5	2,3,4,5	3,4	2,3,4,5,6	2,3,4
37	collisions, kinetic energy, potential energy	2,3,6,7,8	2,3,7,8,9	2,3,6,7	3,4,6	2,3,4,7,8,9	2,3,5,6
38	Hooke's law, friction, Kinetic energy, collisions	2,4	2,4	2,4	3,4	2,5	2,4
39	collisions , kinetic energy, gravitational potential energy,	2,3,4	2,3,4	2,3,4	3,4	2,3,4,5	2,3,4
40	reference frames, energy, oscillations	3,4,6,7,10	3,4,7,8,11	3,4,6,8,11	3,4,5,6	3,4,5,7,8,11	3,4,5,8

Schwarz/Ertel/MSC Software
Interactive Physics Workbook Simulations CD, Second Edition
CD License Agreement

Pearson Prentice Hall
Pearson Education, Inc.
Upper Saddle River, NJ 07458

YOU SHOULD CAREFULLY READ THE TERMS AND CONDITIONS BEFORE USING THE CDROM PACKAGE. USING THIS CD-ROM PACKAGE INDICATES YOUR ACCEPTANCE OF THESE TERMS AND CONDITIONS.

Prentice-Hall, Inc. provides this program and licenses its use. You assume responsibility for the selection of the program to achieve your intended results, and for the installation, use, and results obtained from the program. This license extends only to use of the program in the United States or countries in which the program is marketed by authorized distributors.

LICENSE GRANT
You hereby accept a nonexclusive, nontransferable, permanent license to install and use the program ON A SINGLE COMPUTER at any given time. You may copy the program solely for backup or archival purposes in support of your use of the program on the single computer. You may not modify, translate, disassemble, decompile, or reverse engineer the program, in whole or in part.

TERM
The License is effective until terminated. Prentice-Hall, Inc. reserves the right to terminate this License automatically if any provision of the License is violated. You may terminate the License at any time. To terminate this License, you must return the program, including documentation, along with a written warranty stating that all copies in your possession have been returned or destroyed.

LIMITED WARRANTY
THE PROGRAM IS PROVIDED "AS IS" WITHOUT WARRANTY OF ANY KIND, EITHER EXPRESSED OR IMPLIED, INCLUDING, BUT NOT LIMITED TO, THE IMPLIED WARRANTIES OR MERCHANTABILITY AND FITNESS FOR A PARTICULAR PURPOSE. THE ENTIRE RISK AS TO THE QUALITY AND PERFORMANCE OF THE PROGRAM IS WITH YOU. SHOULD THE PROGRAM PROVE DEFECTIVE, YOU (AND NOT PRENTICE-HALL, INC. OR ANY AUTHORIZED DEALER) ASSUME THE ENTIRE COST OF ALL NECESSARY SERVICING, REPAIR, OR CORRECTION. NO ORAL OR WRITTEN INFORMATION OR ADVICE GIVEN BY PRENTICE HALL, INC., ITS DEALERS, DISTRIBUTORS, OR AGENTS SHALL CREATE A WARRANTY OR INCREASE THE SCOPE OF THIS WARRANTY.

SOME STATES DO NOT ALLOW THE EXCLUSION OF IMPLIED WARRANTIES, SO THE ABOVE EXCLUSION MAY NOT APPLY TO YOU. THIS WARRANTY GIVES YOU SPECIFIC LEGAL RIGHTS AND YOU MAY ALSO HAVE OTHER LEGAL RIGHTS THAT VARY FROM STATE TO STATE.

Prentice-Hall, Inc. does not warrant that the functions contained in the program will meet your requirements or that the operation of the program will be uninterrupted or error-free.

However, Prentice-Hall, Inc. warrants the CDROM(s) on which the program is furnished to be free from defects in material and workmanship under normal use for a period of ninety (90) days from the date of delivery to you as evidenced by a copy of your receipt.

The program should not be relied on as the sole basis to solve a problem whose incorrect solution could result in injury to person or property. If the program is employed in such a manner, it is at the user's own risk and Prentice-Hall, Inc. explicitly disclaims all liability for such misuse.

LIMITATION OF REMEDIES
Prentice-Hall, Inc.'s entire liability and your exclusive remedy shall be:
1. the replacement of any CD-ROM not meeting Prentice-Hall, Inc.'s "LIMITED WARRANTY" and that is returned to Prentice-Hall, or
2. if Prentice-Hall is unable to deliver a replacement CD-ROM that is free of defects in materials or workmanship, you may terminate this agreement by returning the program.

IN NO EVENT WILL PRENTICE-HALL, INC. BE LIABLE TO YOU FOR ANY DAMAGES, INCLUDING ANY LOST PROFITS, LOST SAVINGS, OR OTHER INCIDENTAL OR CONSEQUENTIAL DAMAGES ARISING OUT OF THE USE OR INABILITY TO USE SUCH PROGRAM EVEN IF PRENTICE-HALL, INC. OR AN AUTHORIZED DISTRIBUTOR HAS BEEN ADVISED OF THE POSSIBILITY OF SUCH DAMAGES, OR FOR ANY CLAIM BY ANY OTHER PARTY.

SOME STATES DO NOT ALLOW FOR THE LIMITATION OR EXCLUSION OF LIABILITY FOR INCIDENTAL OR CONSEQUENTIAL DAMAGES, SO THE ABOVE LIMITATION OR EXCLUSION MAY NOT APPLY TO YOU.

GENERAL
You may not sublicense, assign, or transfer the license of the program. Any attempt to sublicense, assign or transfer any of the rights, duties, or obligations hereunder is void.

This Agreement will be governed by the laws of the State of New York.

Should you have any questions concerning this Agreement, you may contact Prentice-Hall, Inc. by writing to:
ESM Media Development
Higher Education Division
Prentice-Hall, Inc.
1 Lake Street
Upper Saddle River, NJ 07458

Should you have any questions concerning technical support, you may write to:
New Media Production
Higher Education Division
Prentice-Hall, Inc.
1 Lake Street
Upper Saddle River, NJ 07458

YOU ACKNOWLEDGE THAT YOU HAVE READ THIS AGREEMENT, UNDERSTAND IT, AND AGREE TO BE BOUND BY ITS TERMS AND CONDITIONS. YOU FURTHER AGREE THAT IT IS THE COMPLETE AND EXCLUSIVE STATEMENT OF THE AGREEMENT BETWEEN US THAT SUPERSEDES ANY PROPOSAL OR PRIOR AGREEMENT, ORAL OR WRITTEN, AND ANY OTHER COMMUNICATIONS BETWEEN US RELATING TO THE SUBJECT MATTER OF THIS AGREEMENT.